# WIDE AWAKE AT 3:00 A.M.

# WIDE AWAKE AT 3:00 A.M.
## By Choice or by Chance?

### RICHARD M. COLEMAN

Foreword by
William C. Dement, M.D., Ph.D.
President, Association of Sleep Disorders Centers
and Clinical Sleep Society
Director, Stanford Sleep Disorders Research Center and Clinic

W. H. Freeman and Company
New York

*Paperback cover illustration by Roy Wiemann.*

This book was published originally as a volume of
*The Portable Stanford*, a book series published by the
Stanford Alumni Association, Stanford, California.

**Library of Congress Cataloging in Publication Data**

Coleman, Richard M. (Richard Mark).
   Wide Awake at 3:00 A.M.

   Bibliography: p.
   Includes index.
   1. Circadian rhythms.   I. Title.
QP84.6.C65   1986      612'.022      86-7665
ISBN 0-7167-1795-6
ISBN 0-7167-1796-4 (pbk.)

Printed in the United States of America

1 2 3 4 5 6 7 8 9 0 ML 4 3 2 1 0 8 9 8 7 6

# Contents

# Foreword

D r. Coleman has written a book that is a major contribution to a very important field, and I am more than pleased to write this foreword. Not only is the topic of sleep and alertness a timely topic, but Dr. Coleman's book takes its place alongside the writings, research, and lectures that have established Stanford University as a preeminent source of useful knowledge about sleep. Richard Coleman gained considerable experience and made a significant contribution as the director of the Stanford University Sleep Disorders Clinic at a crucial time in its history, when it was stimulating an expanding interest in the field and establishing sleep disorders as a nationwide discipline. He also served as chairman of the Multicenter Case Series Committee for the Association of Sleep Disorders Centers and compiled and published data on 8,000 cases diagnosed and treated in eleven clinics across the United States.

In this book Dr. Coleman has dealt with areas of human health that are associated with the scheduling of sleep and wakefulness. In an ideal world there would presumably be a normal schedule that we would adhere to: We would sleep when we were sleepy; this would presumably be in the hours of darkness. We would sleep in a safe place, so we would have no anxiety. During the daytime we would be fully alert, without the need for caffeine or other stimulants, and possibly planning a daily nap. Almost no one in the world now actually enjoys these ideal conditions. Increasing numbers must follow abnormal schedules and get too little sleep. Nearly half of us must stay awake at night to watch over the other half as they sleep.

With each turn of the world, nearly five billion people go through "the valley of the shadow of death." Why are we wide awake at 3:00 A.M.? Some because their jobs demand it; but not all who are scheduled to be wide awake at 3:00 A.M. can remain awake, and this threatens the safety of all. Others because they cannot sleep; the restorative function that sleep should perform is denied them. Dr. Coleman examines these questions and the related issues of biological rhythms, jet lag, and shiftwork scheduling in detail and offers practical guidelines to follow.

Our knowledge about human behavior is constantly expanding, but we have a new world to explore—the world of sleep and wakefulness, schedules and rhythms—and those who work in this area have the right to be called pioneers, even today. Human beings look for new areas to conquer, and in this book Richard Coleman gives us the current maps and beachheads of one of man's last unexplored frontiers.

William C. Dement, M.D., Ph.D.
*Professor, Psychiatry and Behavioral Sciences, Stanford University*
*President, Association of Sleep Disorders Centers*
*and Clinical Sleep Society*
*Director, Stanford Sleep Disorders Research Center and Clinic*

Stanford
1986

# Preface

Several chance events led to my interest in sleep research. While in high school, I was vaguely aware that my sister was spending the summer working in a sleep and dream laboratory, mainly because she tried to enlist me in her research project. I was to wear red goggles during the day to see what effect they would have on my dreams. It sounded strange but interesting, and I might have done it if I hadn't already found a summer job.

A few years later I was completing a college course in physiological psychology and wondering what topic to select for my research paper. One day, as I was walking across the Lake Forest College campus, the wind from off Lake Michigan blew a paper across my feet. I picked it up and saw that it was a report on the physical effects of transcendental meditation. Its statement that some TM practitioners actually fall asleep during meditation reminded me of something my sister had told me—that sleep could be measured. It occurred to me that I had hit on a good research topic. My professor helped me hook up an EEG (polygraph) to a college student willing to have her brain waves recorded while she was doing TM. This first experience fueled my interest: Here was a good way to find out what was going on during different states of consciousness.

In 1973 I was offered an opportunity to work as a night sleep technician at the University of Chicago Sleep and Dream Laboratory, where twenty years earlier REM sleep had been discovered. This time I didn't turn down the offer, and after exploring the world of dreams, I moved on to graduate school in New York and an apprenticeship with two outstanding scientists, Dr. Edward Tauber and Dr. Elliot Weitzman. It was here that my interest in sleep disorders began, an interest that was nourished by the annual sleep research meetings to which my mentors sent me.

My luck held, and one summer, while attending a sleep meeting in Cincinnati, I started a conversation in the hotel swimming pool with a young man—a sleep researcher from Stanford, Dr. Charles Czeisler. It was Chuck who introduced me to the world of chronobiology and time-free environments. We became good friends, and Chuck invited me to give a lecture at Stanford, where I met the Stanford group, the leading sleep research team in the world. Upon completion of my graduate studies I had the good fortune to be invited to work at the Stanford University Sleep Clinic, where Dr. William Dement, who urged me to move to California to continue my studies at Stanford, and Dr. Christian Guilleminault have guided me over the last ten years through the expanding field of sleep disorders.

Because I have always been interested in the practical application of knowledge, it seemed only natural to apply some of my expertise not only to individual patients but also to industry and to sports teams. I hope this book will offer ideas of practical value to shiftworkers, plant managers, jet travelers, and air crews, as well as to persons suffering from sleep disorders, and to the generally health-conscious reader.

There are many people to thank for helping me complete this book: my wife, Melanie, who freed me from child care during the summer vacation when I started writing (I never could have done it without the peaceful month in Martha's Vineyard she provided); my children—Adam, Leah, and Miriam—who endured my weekend work and consequent neglect of them; Sam and Helen Wells for their emotional support; my editor, Miriam Miller, and her assistants, Susan Krever and Gayle Hemenway, for their enthusiastic support and assistance. Finally, I thank my parents, who, when I was in grammar school, helped me rewrite my English composition every Sunday night, and my grandmother, who would ask me at the end of every day, "Well, did you *accomplish* today?"

Richard M. Coleman

Stanford
1986

# 1

# Biological Clocks

At 1:50 A.M., Burlington Northern freight train Number 7843 pulled out of Eagle Butte Coal Mine in Gillette, Wyoming. The engineer and head brakeman operated the 115-car train from the lead locomotive; the conductor and rear brakeman were in the caboose.

At 4:18 A.M. Number 7843 was moving east on the single main track at the restricted speed of 35 MPH. Flashing yellow signals indicated that several trains were on the same main track, about twenty miles ahead at Pedro, Wyoming, but the engineer and head brakeman had nodded off and missed them. At Y.T. Hill, a steep downgrade a few miles before Pedro, the train's rapid acceleration alarmed the conductor in the caboose. He radioed the engineer in the lead locomotive. The engineer woke up and saw the speedometer registering 62 MPH. His application of the air brake startled the head brakeman, who awoke to see the engineer in a state of panic. The engineer asked the brakeman if he had seen the last signal. No. He too had been asleep.

Thirty seconds later, a stop signal came into view, but too late—the train was going too fast. They saw two other freight trains dead ahead at Pedro. The engineer shouted over the radio, "Get off your way car," and sounded his whistle repeatedly. Crew members on the nearest of the two trains at Pedro heard a garbled message but were unaware of an oncoming train.

Five hundred feet before Pedro, the engineer and head brakeman jumped off Number 7843, by now speeding out of control at 45 MPH.

At 4:45 A.M., Number 7843 struck Number 8112. Two crew members in the caboose of 8112 were immediately killed. Only nine days earlier, two other Burlington Northern freight trains had collided head on at 3:55 A.M.; five crew members were killed. The National Transportation Safety Board investigated the accidents and determined as the probable cause the employees' falling asleep and thus failing to comply with restrictive signals.

Both accidents took place shortly after 3:00 A.M., the lowest point in the body's internal cycle of alertness. Because an employee is scheduled to perform at a certain time does not necessarily mean that he will be able to do so. Over the centuries humans, like most other species, have developed a biological timing system—a biological clock or a group of brain cells—that determines when you feel sleepy, when you feel alert, when certain hormones are released, and much more. Unlike the pacemaker in your heart, which you can feel beating every second, the biological clock cycles about once a day. That's why it is hard to measure and why scientists didn't discover its existence until a short time ago. Our internal clock may be an Achilles heel in our round-the-clock society or it may be an unutilized resource that we can harness to improve our health, safety, and performance.

Daily fluctuations in performance and alertness are largely controlled by biological clocks. Human performance cycles and their effect on safety have been a growing concern in industries that must function around the clock, such as chemical plants, aviation, the armed forces, and nuclear power plants. The Three Mile Island accident, largely attributable to human error, also occurred shortly after 3:00 A.M., with a crew that had just started rotating back onto night shift duty. While most of our population is trying to sleep at 3:00 A.M., an important group is struggling to maintain alertness in order to provide security and safety for the rest of us.

It may be difficult to see or feel our biological clock, but we are all under its influence. Usually our internal clock is synchronized with the clock on the wall and everything runs smoothly. For example, when the clock on the wall reads midnight, our inner clock is usually making us feel sleepy, and we are ready for bed. But what happens if your sleep schedule is midnight to 7:00 A.M. but your internal clock is set for peak alertness at 3:00 A.M.? Or your internal clock is set for sleep and you are supposed to be working at 3:00

A.M.? If you could schedule a job interview for any time you wanted, what hour would you choose? What times would you avoid? If you were going to work an 8-hour shift monitoring radar for the Strategic Air Command, what hours should you be on duty to ensure maximum alertness? As we come to understand more about biological clocks it may be possible to solve the problems of insomnia, shiftwork, jet lag, and diminished daytime alertness—problems that affect an increasingly large number in our population.

## Nature's Clocks

Over 2000 years ago, the rhythmic nature of certain biological patterns was observed. Aristotle observed the swelling of the ovaries of sea urchins at full moon; Hippocrates noted daily (24-hour) fluctuations in his patients' symptoms; and Herophilus of Alexandria observed daily changes in pulse rate. A more common observation from ancient times was that certain "sensitive plants" opened their leaves during daylight hours but closed their leaves during the night. A Greek myth described these movements as a continual expression of love for the Greek sun god, Helios. The rhythm implied that the "sleep movements" of these plants, which we now call heliotropes, were a passive response to the disappearance and reappearance of sunlight.

What caused this rhythmic behavior? Despite numerous observations of heliotropes, it was not until 1729 that a French astronomer, Jean Jacques d'Ortous de Mairan, designed a study to answer this question. Previous writers had assumed the obvious—that sunlight caused the leaves and stems to open and the absence of light made them fold. De Mairan conducted a simple experiment: He placed the plant in a dark closet into which no light could enter and peeked inside at various times of the day and night. To his surprise, when placed in total darkness, the leaves remained open during daytime hours. And when he observed the plant during nighttime hours, he found the leaves closed. De Mairan, a busy astronomer, was not interested in pursuing this curious phenomenon. His colleague, Marchant, published a one-page abstract—which was largely forgotten—in the proceedings of the Royal Academy of Paris. This study was the first demonstration of the existence of a *biological clock, an internal (endogenous) mechanism capable of measuring time in a living organism.* Even when isolated from sunlight cues, the plant itself was able to

De Mairan's experiment demonstrated the existence of a biological clock in plants.

tell whether it was daytime or nighttime. This fact, however, did not gain acceptance among scientists until 250 years later.

A few researchers expanded upon de Mairan's observations. Henri-Louis Duhamel (1758) thought that perhaps some light had leaked into de Mairan's dark place; he therefore observed plants in a dark, underground wine cellar. He went even further—he put the plant in a trunk, covered it with blankets, and placed the trunk in a closet. He too found that the plants could keep track of time and that temperature changes were not responsible for plant sleep movements.

Another botanist, de Candolle (1832), kept his plants in constant daylight by placing them near a series of lamps burning at a steady intensity. Not only did the plants continue to open and close their leaves regularly, but the daily cycle of plant movement was 22 hours. It takes 24 hours for the earth to make one complete rotation on its axis, which is equivalent to one day or one light-dark cycle, so that the clocks of these plants ran fast in respect to the 24-hour day of the earth's rotation. This was the first demonstration of a phenomenon called *free-running:* When freed from the constraints of the usual 24-hour time cues provided by sunlight and night darkness, the plant ran its own independent day length.

These studies demonstrated several major principles of biological clocks: (1) They are internal physiological systems that can measure the passage of time; they are not passive responders to environmental cues. (2) They have their own daily cycle length, which is close to, but not exactly, 24 hours. (3) When exposed to normal environmental cues, such as the day-night cycle, the organism can adapt to a 24-hour day. (4) When free of the normal day-night cycle, the organism's own internal cycle length determines its behavior.

Biological clocks give organisms the ability to measure time. In an environment with fluctuating light and dark, temperature, day length—all consequences of the earth's rotation—these biological clocks allow organisms to predict and anticipate major changes in environmental conditions. For example, bees can measure the passage of time so that they can arrive at a specific flower when it is releasing pollen; animals can anticipate predator activity and changes in food supply, so that they can activate themselves at critical times, or put themselves to sleep at times when waking would be maladaptive.

Biological clocks organize our society. Without them, our behavior is chaotic. In infants, for example, the biological clock is poorly developed until three to four months after birth. During the first three months, the infant's feeding and sleeping schedule is random, causing considerable disruption for the mother and family. In fact, mothers spend a great deal of time worrying about their infants' sleep schedules, as well as their own. As the *circadian* (daily cycle) system matures, infants develop a long nocturnal sleep period, much to the family's relief. This allows the child to synchronize with the sleep-wake schedule of the rest of the family. Case studies of adults with lesions of the brain area housing the biological clock include reports of erratic, random sleep-wake schedules—although they may sleep about 8 hours out of every 24.

In the first half of the twentieth century, a variety of studies of single-cell organisms and of bees, cockroaches, birds, squirrels, and monkeys were undertaken. All of these species were found to have internal timing systems, or biological clocks, that controlled the timing of behavior—from eating and drinking rhythms to cycles of growth and cycles of birth. Stanford Professor of Biology Colin Pittendrigh in a number of elegant studies of successive generations of fruit flies *(Drosophila)* demonstrated that circadian clocks are innate and genetically transmitted rather than learned behaviors. More recently a series of investigations have confirmed the presence of biological clocks in humans.

## A World Without Time

In order to find out more about the human biological clock, scientists had to construct special time-free environments. Volunteers would have to be isolated from time cues or *zeitgebers* (literally "time givers") to determine whether the observed rhythms in humans are passive responses to the environment or true biological clocks. Several hundred human volunteers have been studied in this manner, some in underground caves in the Swiss Alps, some in Germany—in underground bunkers shielded from electromagnetic fields—and some in special apartments at Stanford University and in the Bronx in New York City. The human volunteers were studied in a world without time—without clocks, wristwatches, windows, radios, TVs, or telephones—just as de Mairan's sensitive plants were.

When Dr. Elliot Weitzman and Dr. Charles Czeisler organized the opening of the Laboratory of Human Chronophysiology at Montefiore Hospital in the Bronx, I was fortunate enough to be there, as were a number of other sleep researchers from Stanford University. On the fifth floor in one of the old hospital wings, three specially soundproofed rooms were constructed—two one-bedroom suites and a control room in the middle. The bedrooms had no clocks and no windows. Volunteers were recruited to live in these apartments for anywhere from one to six months. Big signs were posted to keep housekeeping, nursing, and engineering staffs out of this restricted area. To ensure that we were working with "normal" volunteers, each subject was carefully screened.

As a psychologist, one of my tasks was to find someone crazy enough to live in this environment who could still be considered normal. Over a few years a variety of subjects whose ages ranged from twenty to eighty were evaluated—students, craftsmen, newspaper reporters, artists. Any "stable" person who had a good reason for wanting to "get away from the world" for a few months was considered a good candidate. The volunteers also had the advantage of making several hundred dollars a week and being simultaneously deprived of the temptation to spend it. One subject who lived in the apartment for six months not only received credit from his university for independent study but made up several "incompletes"—all while enjoying free room and board. (Perhaps not surprisingly, most subjects did not finish all the grand tasks and goals they had anticipated they would, even given their freedom from social duties and distractions.)

While in the apartment, volunteers were free to interact with the staff, which was far more numerous than the subject group (it can take up to twenty people to run one subject). The technicians and doctors were on a totally random work schedule, generated by a computer, so that the subjects could not estimate the time of day by noting who was working a particular shift. The staff was not allowed to wear wristwatches, or to say "good morning" or "good night," but only "hello," and male staff were instructed to shave before entering the apartment to eliminate any trace of "five o'clock shadow." Subjects received only outdated magazines and newspapers, and radios and TVs were not allowed. Within these constraints subjects were free to plan their own routines: reading, working, exercising, playing music.

In short, everything was done to make the volunteers' stay as pleasant as possible but without giving them any hint of the time of day. The major time-givers (zeitgebers) for humans—clocks, social cues, and sunlight-nightfall cycles—were carefully eliminated. While the subjects lived in the apartments, the technicians in the control room took a series of continuous physiological measurements: tiny, frequent blood samples (every twenty minutes) to evaluate hormones; room, skin, and rectal temperature recordings every minute; performance and alertness tests nearly every hour; and brain-wave monitoring of every sleep episode. Although a tremendous amount of data was collected, the subjects became accustomed to the recording devices and pursued an essentially normal indoor life.

The actual experiment was rather simple in comparison to the elaborate preparatory efforts made to ensure a reliable study. For the first twenty days and nights, volunteers were instructed to get into bed and try to sleep when instructed to do so. At this early phase of the study the experimenters in the control room (who knew what time it was) turned the lights off at midnight. At 8:00 A.M. the volunteers were awakened, the lights were turned on, and they were free to follow their own daytime routine until the following midnight, when the next sleep period was scheduled.

After twenty days, the experimenters made a significant change in their instructions: "From now on you are free to go to sleep whenever you like and wake up whenever you like. Just decide, based upon your own feelings, when you are ready for sleep and when you are ready to wake up." Subjects were instructed to avoid taking naps and to aim for one sleep period per day.

What kind of sleep schedule would you maintain in this situation? Surprisingly, the results for nearly all subjects were identical. The first night they fell asleep about an hour later than they had during the previous twenty days and slept about an hour later, so that their sleep time shifted from 12:00 P.M–8:00 A.M. to 1:00 A.M.–9:00 A.M. The following night they delayed their sleep schedule to 2:00 A.M.–10:00 A.M., and this pattern continued: They delayed their sleep period by one hour until, for example, after several weeks into the study, they stumbled upon 4:00 P.M.–12:00 P.M. as their sleep time. Upon awakening at midnight they would buzz the technician over the intercom and place their order for breakfast. The technician on night shift (who was hoping the subjects would sleep through the end of his shift at 2:00 A.M.) scrambled up some eggs and served breakfast at about 12:30 A.M.

Meanwhile, over their 1:00 A.M. breakfast and an outdated *New York Times*, the volunteers planned their workday, opting to go to sleep next at 5:00 P.M. and continuing to drift around the clock. Most reported that they felt more alert during the study than they had in their usual routine in the normal "timed" society. The results of these sleep-wake cycles showed that most subjects averaged a 25-hour day—that is, left on their own, free of time cues, humans have an internal day length of 25 hours. Not only did the sleep-wake cycle

FIGURE 1-1

## Free-running

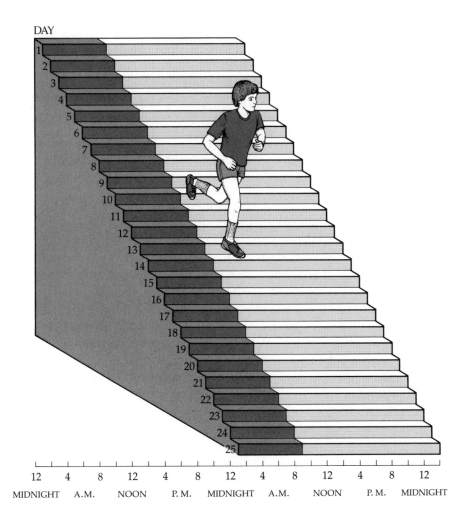

Free of time cues, humans choose to go to sleep one hour later each day. This natural tendency to later hours is called free-running.

show this pattern, but body temperature, hormones, performance—all showed evidence of drifting to a longer than 24-hour day. This tendency to drift naturally to later hours is called "free-running." When living on the 25-hour day, subjects slightly increased their total sleep time to about 8½ hours, maintaining the same ratio of sleep to waking as they normally did in a 24-hour day. They slept roughly one-third of the time during both the 24-hour day (scheduled-study phase) and the 25-hour day (free-running phase). Having replaced de Mairan's sensitive plants with human volunteers, we had achieved similar results: They had established their own biological day length while living in a time-free environment (see Figure 1-1).

Here is an excerpt from the journal of Preston Keogh, one of the subjects who served time in the Bronx apartment:

> When I was out of college I was broke and this was a way of making some money. It was virtually impossible to spend money in the apartment but I figured out a way. I got the *Wall Street Journal* delivered. Although they gave me out-dated copies and I couldn't play the stock market, I spent a lot of money on mail order services.
>
> I spent a lot of time reading and writing to make up some incompletes. I got more done in a month than I normally did in a whole semester. I thought it was important to have a certain routine to maintain a measure of sanity, so I wore a shirt and tie, and shaved every day. One of my biggest problems was that my pants were wool and I couldn't get the creases pressed. So sometimes I walked around with a shirt and tie and shorts!
>
> They told me they wanted to create a normal environment and then they banned drugs, sex, and alcohol. When the study was completed I said I'd be happy to volunteer again with these three items and they could monitor me all they wanted. They still haven't called me back.
>
> Sometimes I felt like a prisoner, trading my youth for money. Although I didn't feel crazy, I thought others might think I was. I'm quite comfortable with myself a little con-fined. I was happy as a clam. I could tell they were also a little strange, more interested in my urine samples than in some fascinating dreams.

They took blood samples every fifteen minutes. I had a catheter in my arm, and a butt probe and all these things were attached to a movable pole. The first few days there was a definite presence but after the first week it became a part of you. It was like having a tail.

They told me they wanted to figure out the ideal day length for scheduling submarines and spacecraft. They were very vague and said they'd tell me afterwards. During the study I didn't worry about it. I had never gotten anyone else to pay me to go to sleep and I was getting quite good at it. They told me not to try to figure out what time it was and that their staff was on a random schedule set by a computer. I decided early on not to figure out what time it was, so I didn't try. I never knew what time it was and didn't worry about it, except one time a technician came in with tuna fish on his breath and bloodshot eyes. I said, "Pretty tough night, eh?"

I didn't feel isolated from people. They could have had a few more women on their staff. Sometimes they played chess with me but I always beat them so they lost interest in playing with me. When I first got out an old girlfriend showed up for the getting-out party. It was disorienting to be able to see the distance. I could see ten miles and I had been confined to forty feet! We drove through downtown New York, got a haircut and did some things that were long overdue. The experimenter told me about biological clocks, and I left with five or six thousand dollars. I was the longest one ever done in North America—102 days—until some wacko guy lived there for six months!

Why is our biological day 25 hours? Why does the human internal pacemaker, our biological clock, operating on a 25-hour day, run a little slow in comparison to the 24-hour rotation cycle of the earth? There are no definite answers. Each species studied appears to have a specific "biological day": Flying squirrels have a 23-hour day, monkeys a 24½-hour day. Perhaps humans have a selective advantage; our 25-hour clock makes it necessary for us to develop an adaptability to changing day lengths caused by changes in the seasons or in the amount of daylight. A species with an exact 24-hour clock, incapable of adjusting or resetting, could be at a disadvantage during the changes of daylight in winter or summer.

Even though our natural biological day is 25 hours, our biological clocks respond to time cues from the outside world. We can therefore prevent ourselves from drifting around the clock by synchronizing ourselves to the 24-hour day with a regular sleep or work schedule, alarm clocks, daylight, or other stable routines. If the time cues in the environment are irregular, such as those experienced by shift-workers or jet travelers, then our biological rhythms can free-run, as if we were living in the time-free apartment in the Bronx.

Let us pause for a moment to define some terms and to report the findings of studies of biological rhythm. *Chronobiology* is that branch of science that studies rhythms of life that are an outgrowth of biological systems. Chronobiology is based upon repetitive measurement of naturally occurring physiological phenomena. The most widely studied biological rhythms are *circadian* rhythms (*circa*, from Latin, meaning "about;" *dian*, from the Latin *dies*, for "days")—cycles that fluctuate on a daily basis. *Infradian* cycles, those lasting a month or longer, include the menstrual cycle in adult females, or the seasonal *circannual* cycle (about one year) of hibernation in certain mammals. Many of the infradian cycles are difficult to measure because a complete cycle may last up to several years. *Ultradian* rhythms, such as heart rate or respiration, have cycles much shorter than 24 hours. These are *biological rhythms*, which should not be confused with *biorhythms*.

The study of biological rhythms is to the study of biorhythms as astronomy is to astrology. Biorhythm is not a science; it is the practice of forecasting events in the life of a single person based primarily on the date and time of the individual's birth. Biorhythm theory postulates a 33-day intellectual cycle, a 28-day emotional cycle, and a 23-day physical-strength cycle. The quality of performance that can be anticipated depends, according to this theory, upon the interaction of these cycles. Researchers using scientific methodology to investigate the occurrence of accidents, deaths, and the results of athletic contests have found that the timing of these events did not correspond to the predictions of biorhythm practitioners.

Even in the 1960s a few scientists continued to argue that de Mairan's phenomenon was not evidence of an innate biological timing system. True, plants maintained their rhythm even though light did not penetrate the plants in their dark quarters, but perhaps, they suggested, some subtle environmental stimuli caused by the earth's rotation (such as electromagnetic fields or cosmic radiation) served as cues by which plants and animals could sense the time. To eliminate these potential subtle effects, researchers at the South Pole placed hamsters, fruit flies, and plants on a table rotating counter to the earth's rotation. All continued to show the same rhythmic behavior that de Mairan first described. To date, there is little evidence to indicate that a geophysical force is giving subtle time signals, whereas many studies have demonstrated a genetic inheritance of biological clocks. Future time-isolation studies in space will undoubtedly provide the final proof for existence of endogenous (internal) biological clocks.

# 2

# Setting Your Clock

He gets up at 4:45 A.M. to beat the rush hour commute. By 5:30 he has exercised, showered, had his breakfast, and is on the road. He's a business executive whose peak time is 8:30 A.M. She's a newscaster who anchors the 11:00 P.M. news; she doesn't get to bed until 3:00 A.M., sleeps until noon, and leaves for work at 4:00 P.M. She's an owl and he's a lark. They look forward to the weekends, when they can be together.

Only about 10 percent of the population are extreme owls or extreme larks. Most of us function somewhere in the middle.* We can get up early and function for a few days or stay up late and function well for a few nights. Work schedules and sleep schedules are the primary forces that make an individual into an extreme owl or lark. Once you become accustomed to a late sleep schedule, all your biological cycles will synchronize to these new terms.

All combinations of larks and owls, whether as marriage partners, roommates, or friends, raise thorny questions. Larks see owls as lazy; owls see larks as party poopers. Yet each is merely trying to arrange life to accord with times of best performance. Owls will stay up late to cram for a final exam the next day; larks will get up early on the morning of the test. Let's take a closer look at the relation between biological clocks and performance.

That sleep patterns govern performance in a simple and direct way—that is, we think we perform badly because we have had too little sleep—is a common misconception. The relationship between

*See Appendix for Owl and Lark self-test.

sleep and performance is not a simple one: The key to performance lies not only in how much or how little one sleeps but also in one's circadian rhythm. Circadian rhythms continue, giving us peaks and valleys of performance, even when we do not sleep at all.

In order to correlate performance with the biological clock, we first need a reliable method for determining what time it is in the body. One of the first biological rhythms to be carefully studied in humans was the daily fluctuation of core body temperature. As far back as 1778, J. Hunter reported that a drop of 1.5 degrees in body temperature accompanied sleep. This finding is open to two interpretations: That a lack of activity makes the body temperature fall during sleep, or that an internal clock causes the temperature to wax and wane. The answer was not found until the mid-twentieth century, when investigators devised more complex experiments that enabled them to study the effect of exercise, alcohol, activity, and reversed sleep-wake schedules on the temperature cycle. Dr. Nathaniel Kleitman, the father of modern sleep research, carried out many of these studies. He found that even in a subject who is kept awake all night, but resting in bed, the body temperature will continue to drop. He also found that if a person is kept awake at night but allowed to sleep during the day, his temperature rhythm will eventually invert to accommodate the new schedule: It will peak during the night and fall during daytime sleep. These studies suggested that a human biological clock controls the temperature cycle and that this clock can be reset.

More recently, a certain cluster of neurons has been identified as the main generator of the human biological clock. At present, we cannot gain direct access to this clock mechanism in order to discover what time it is in the body. However, by looking at the output of the clock, such as the temperature rhythm—which is easy to measure—or the rhythms of various hormones, we can tell biological time. A rough analog to this is, of course, telling time by looking at the hands of the clock on the wall but without understanding or seeing the ticking mechanism behind the clock. For example, when your body temperature is at its highest point for the day, you know that your biological clock is set at 12:00 circadian time. In most individuals active during the day and sleepy at night this would occur during the late afternoon.

Unfortunately, measuring biological time is a little more complicated than taking a temperature reading in the late afternoon. A hot bath, ice cream, exercise—all these will have a dramatic short-term effect on body temperature. To really see the circadian rhythm (about a day), we must study a longer time period. Several strategies are available for measuring biological time accurately. Researchers who are able to take frequent measures of body temperature over several days can make the most accurate assessment. Remember, since the clock oscillates on a daily basis it would take at least 24 hours to determine an individual pattern. Some researchers advocate measuring other selected physiological variables, such as certain hormones, less likely to be influenced by those external factors that influence body temperature.

Scientists are currently investigating all these possibilities. Dr. Laughton Miles and Dr. Elliot Weitzman helped develop an ambulatory recording device that weighs no more than and looks much like a Sony Walkman. Worn on a belt, it is capable of measuring temperature, heart rate, and activity patterns while the subject goes about his daily routine. When plugged into a computer it displays information concerning peaks and troughs of biological time.

Nathaniel Kleitman documented a close relationship between the body temperature cycle, which tells biological time, and fluctuations in performance and alertness. He discovered that as body temperature decreases—at the low point of the biological day—performance and alertness also decrease. In general, there is a daily fluctuation in performance and alertness that correlates with the body temperature cycle. Individuals who stay awake for 24 hours show peak performance near midday and lowest performance from 3:00 A.M. to 5:00 A.M. However, even in subjects who have been kept awake all night, performance starts to improve around 6:00 A.M. Many night workers and all-night partyers are surprised at the renewed vigor they experience at dawn after staying up the previous night. Personnel reading gas meters, answering telephones, adding numbers, doing sonar detection, and responding to warning signals—all sustained, repetitive tasks—show increased error rates at the low point of the body's internal cycle. There are exceptions. Tasks requiring brief bursts of intense concentration are less affected by circadian rhythms. For example, if your house catches fire at 3:00 A.M., you will probably respond with great efficiency and energy. Memorization, too, is best accomplished in the early morning near the temperature minimum. Finally, performance may show a drop at the "post-lunch dip," a phenomenon we will look at in Chapter 8.

FIGURE 2-1

# Circadian Timing
## of Physiological Functions
### Over Two Days

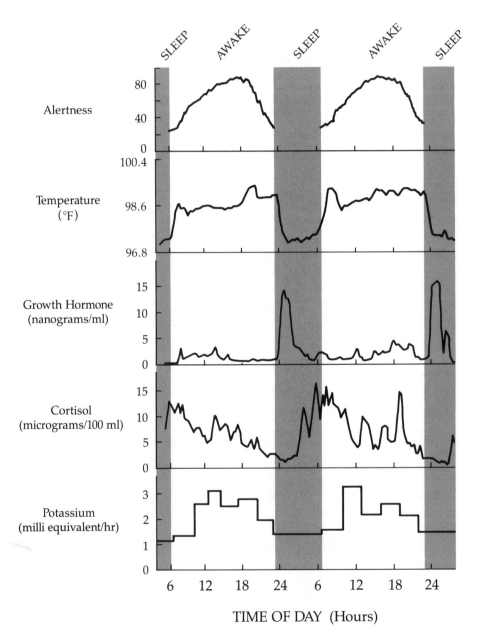

FIGURE 2-2

## Sleep and Alertness Cycle
## of a Worker Starting a New Night-Shift Rotation

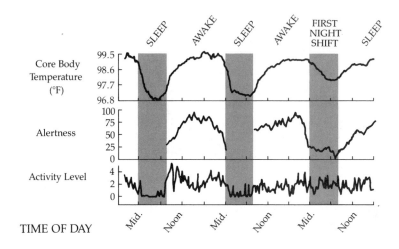

### Our Inner Clocks

If we trace the biological rhythms of a day worker, we can learn something about how these inner clocks work. The rhythms are the output of the internal clock, like the hands of the clock on the wall. Figure 2-1 shows the rhythms of alertness, body temperature, growth hormone, cortisol (a hormone that is secreted to help deal with stress), and potassium excretion for a typical day worker. Each physiological function has its own unique pattern that cycles over a 24-hour period. During the daytime (wake period) body temperature is high, but during the night (sleep period) it drops by two to three degrees. Alertness cycles closely follow the body temperature curve. We feel most alert when our body temperature is highest and sleepiest when it is lowest. Growth hormone, however, shows the opposite pattern, secreted at night but rarely during the day. Cortisol and potassium also have their own patterns, which repeat day after day. The biological clock in the brain controls all these cycles. As long as we stay on a regular day schedule our clock operates efficiently and we hardly notice it at all.

Asking the biological clock to adjust to new schedules, whether as a result of work shifts or of jet travel, can put a real strain on the system. You can end up feeling sleepy when it's time to work, and alert when you go home to try to sleep. Figure 2-2 shows how the biological clock responds in a shiftworker coming onto the first night shift, after being off for two days. Even though the technician is active between midnight and 8:00 A.M., the body clock sets a low temperature pattern, which makes the employee sleepy on the job. When the shiftworker gets into bed at home, around 9:00 A.M., he feels restless, unable to sleep. That's because his body clock is set for being alert and awake at that time. Eventually, the temperature-alertness rhythm will invert so that the operator is alert at midnight but ready to sleep at 9:00 A.M. Despite certain limitations, our inner clocks can adjust to changing schedules.

### Running on Automatic Pilot

One performance measure familiar to most of us is driving. Studies of single-vehicle truck accidents—trucks overturning, jackknifing, going off the road of their own accord, colliding with a fixed object like a bridge abutment—show a 700 percent increase between midnight and 8:00 A.M., with the peak occurring between 3:00 A.M. and 6:00 A.M. This finding is more startling in view of the fact that there are fewer trucks on the road at these hours than at any other time of day. These accidents, cited by the U.S. Bureau of Transportation as likely to be caused by "dozing at the wheel," occur most frequently at the low point of the biological day.

To find out more about night-driving performance, a group of German investigators designed a night-driving experiment. The route consisted of 50 kilometers of country roadway, 100 kilometers of monotonous expressway, and another 50 kilometers of country roadway. For half the test, the driver was on the Autobahn expressway, traveling close to 70 miles per hour. Electrodes were placed on the scalp and near the eyes to measure brain-wave activity and eye blinks during the 200 minutes of night driving. Tape recorders on the back seat kept track of the physiological data. During normal wakefulness the brain waves have a distinctive pattern and eye blinks occur frequently.

In analyzing the data, the scientists found a puzzling 20-minute segment: There were no eye blinks at all; the eyes were either completely shut or wide open. During part of this phase, the scalp electrode recording indicated that the brain was asleep. Given the fact that there was no accident, it can be presumed that the driver was going along at 70 MPH with his eyes open, his brain asleep, and his hands set rigidly at the wheel. This special state of consciousness—in which the brain is asleep but low-level performance continues—is called *automatic behavior*, or *automatic pilot*. During such times operators of equipment are functioning at a low level and are unable to anticipate events. If an alarm—such as a police siren—should sound, the driver probably could wake up and perform well. In many situations, however, it is critical to maintain alertness and anticipate events. (In a follow-up interview the night driver confirmed that he had had difficulty fixating during the same 20-minute interval.)

Automatic behavior is not confined to automobile driving. In a performance study of Swedish locomotive engineers, monitored with brain-wave recordings, 65 percent nodded off during a 5-hour nighttime 212-mile run, while only 13 percent nodded off during the same run in the daytime. The data suggest that it is not monotony that causes automatic behavior but the inability of the biological rhythm of alertness to adjust to night hours. Given this premise, we can assume that the Burlington Northern train crashes described earlier were probable rather than improbable events.

An extreme form of automatic behavior, night-shift paralysis, is reported by nurses. They describe an inability to make large body movements and a sense of paralysis while performing sedentary tasks such as reading or writing. They experience an inability to get up and perform their jobs for several minutes. Nighttime computer program operators have been reported to run the same expensive data analysis program over and over again while in an automatic behavior state, running up a huge bill by the time they become fully awake in the morning! Automatic behavior is also found in certain patients with sleep disorders accompanied by impaired alertness.

### Sleeping on Schedule

Although scientists have not developed a simple technique to measure individual circadian patterns, several general principles are known to apply to the human species. Our biological clock is capable of free-running and resetting, but there are strict limitations. For example, if your sleep-wake schedule is from 11:00 P.M. to 7:00 A.M., it is fairly easy to stay on this routine day after day. Our biological clock can easily adjust to a 24-hour schedule. To prevent the system from free-running to a 25-hour day, we merely set an alarm clock to ring every day at the same time, maintain the same bedtime, and allow ourselves to be awakened by sunlight or children. If you try to get into bed at 10:00 P.M. you may, as most people do, struggle a bit to fall asleep, but you will be able to manage it within a short time. Selecting a bedtime 1 hour earlier than usual is equivalent to living on a 23-hour day, but the biological clock cannot readily adjust to day lengths of 22 or 21 hours. If you try to get into bed at 9:00 P.M. or 8:00 P.M. you will probably be unable to fall asleep. On the other hand, going to bed later than usual implies living on a longer than 24-hour day. Since our natural day length gravitates to 25 hours, it is much easier to stay up later than to go to bed earlier. In general, our 25-hour clock can be reset about 2 hours each day, allowing humans to live comfortably on a 23- to 27-hour day.

Over the weekends, in fact, many Americans allow their sleep-wake cycle to follow the natural drift of their circadian system. By turning off the alarm clock and releasing ourselves from reporting to work at a given hour we eliminate *zeitgebers*, which synchronize us to a 24-hour day; this is equivalent to living in a time-free environment. On Friday, many adults stay up a little later, going to sleep at midnight and sleeping in until 8:00 A.M. On Saturday night they may stay up until 2:00 A.M. and on Sunday sleep in until 10:00 A.M. If our biological clock ran at a 23-hour day, it would be a privilege to go to bed earlier and wake up earlier on the weekends, rather than sleep in!

### Monday Morning Blues

The problem with the pattern of weekend free-running comes in facing the next work week. On Sunday night, many weekend partygoers try to get back to their usual sleep schedule of 11:00 P.M. to 7:00 A.M. In trying to move back their sleep-wake rhythms by 3 hours, they try to live on a 21-hour day. The biological clock cannot easily adjust from its programmed pattern of 25 hours. As a result, Sunday night finds many day workers in bed at conventional bedtimes but unable to fall asleep. Although these same people may realize that it is foolish to go to bed 3 hours early (i.e., at 8:00 P.M.) in the middle of the week, they do not realize that the same standards apply to Sunday nights.

Many insomnia patients report Sunday as their worst night. Some report having initially had only Sunday-night insomnia, which later developed into a full-blown seven-nights-a-week problem. If you're

**What we envision as a hidden clock is actually a group of neurons in the brain that control daily physiological rhythms.**

lying in bed when your internal clock is setting you to be awake, it's not suprising to find your mind very active, worrying about the work you have to face the next day. Furthermore, after worrying for 3 hours about the coming week, it may be hard to fall asleep even when your internal clock has scheduled you for sleep.

Perhaps the most common consequence of changing sleep-wake cycles over the weekend is the "Monday morning blues." Waking up at 7:00 A.M. on Monday morning may be the equivalent of 4:00 A.M. on body time, the sleepiest part of the internal cycle. That complaints of sluggishness and fatigue are greatest on Monday is understandable—workers are waking up on Mondays at an hour for which their bodies had been geared for sleep on the previous two days. The Monday morning blues gradually fade as they re-establish a regular sleep-wake schedule over the next few workdays.

### Sleep Length and Biological Time

Hundreds of human physiological functions exhibit circadian rhythms, ranging from respiratory rate to red-blood-cell count to urine metabolites. Under special circumstances such as shiftwork or jet lag—or in a time-free environment—when time cues are lacking or erratic, human subjects may develop a state called *internal desynchronization*. Various physiological rhythms that are usually linked go out of phase. For example, sleep periods may occur at the maximum point of the body temperature cycle instead of the trough. Working, studying, and performance may occur at the low point of the cycle instead of at the peak. This is likely to be a time of increased safety risk. Studying internal desynchronization has led scientists to discover that the amount of sleep one obtains is largely determined by what time it is in the body. A common misconception is that the length of sleep is determined solely by prior wakefulness; that is, if you miss a night of sleep, you are likely to sleep much longer the following day or night. The results of sleep deprivation studies show otherwise.

In 1965 the Stanford group led by Dr. William Dement studied Randy Gardner, a high-school student trying to break the world record for sustained wakefulness of 260 hours. He achieved his aim, staying awake for 11 days—264 hours—under careful observation. What was surprising was his recovery sleep pattern. Instead of "making up" the lost sleep, Randy finished his vigil at 6:12 A.M., promptly fell asleep and slept for 14 hours and 40 minutes, awakening at 8:52 P.M. He then stayed up for 24 hours and resumed a normal nocturnal sleep pattern of 8 hours a night.[1]

Non-sleep-deprived volunteers undergoing internal desynchronization can easily sleep for 14½ hours if their internal biological clocks are set for it. Sleep periods beginning at the time of high and falling temperature cycles are longest; sleep periods beginning at the time of low and rising temperature cycles are shortest. This is akin to the common experience of staying up for an all-night party and trying to start one's sleep period the next morning. After being awake for 24 hours, most of us expect to have a long sleep, adding several recovery hours to our standard 8 hours. In practice, most individuals report sleeping only 4 to 6 hours of poor-quality sleep under these conditions. The reason for this is that our biological clock is like an alarm clock, signaling us to wake up. All of our body rhythms have been set from previous days to be gearing up for alertness in the morning. The biological system struggles to alert us once again, programming us for the day's activity, actually fighting the "recovery sleep."

*Sleep loss cannot be made up.* There is no direct correlation between hours of sleep lost and hours of sleep made up. The length of sleep is affected by two factors: what time it is in the body and prior sleep deprivation. On a regular synchronized sleep schedule with constant bedtimes and wake times, most adults sleep 7 to 8 hours. However, an individual on a regular 11:00 P.M. to 7:00 A.M. sleep schedule who stays up an entire night and goes to sleep at 7:00 A.M. or 11:00 A.M. the next morning in a quiet, dark room (i.e., after 24 or 28 hours of sleep deprivation) will sleep only 4 to 5 hours; that is, he will actually sleep less following a night of sleep deprivation, because the internal alarm clock is pre-set! If that same person were to remain awake until 7:00 P.M., after 36 hours of sleep deprivation, he would probably sleep 10 hours—until 5:00 A.M. the following morning—because he has started his sleep after the peak of his biological day has passed.

### Light-Dark Cycles

S unlight seems to affect our daily moods. In the Arctic Circle "big eye" (insomnia) and "arctic hysteria," a mood disorder, are marked during midwinter, when there is virtually continual darkness for 24 hours, and suicide rates are highest in December and January. These symptoms are not generic to Eskimo populations; they occur in new arrivals (non-natives) after they have lived there for a year. The lack of direct sunlight may not only be psychologically upsetting; it may also upset the delicate physiology of biological pacemakers.

In the subarctic latitudes, sunlight deprivation may occur in modern office buildings, in large communication and power control stations, in nursing homes, prisons, and intensive care units, and among night-shift workers. For some individuals, seasonal affective disorder (SAD) occurs as the days grow shorter. The symptoms include depression, withdrawal from social activity, sleepiness, and increased weight.

This disorder may be related to an abnormal circadian rhythm of melatonin secretion. Under normal conditions, the pineal gland secretes the hormone melatonin at night and not during the daytime. Sunlight and bright room light (about 2000 lux, or ten times normal indoor room light) are effective in suppressing nighttime melatonin secretion. By exposing these SAD patients to intensive bright light for 3 to 5 hours per day (patients actually sit close to bright fluorescent light), researchers were able to simulate a 13-hour spring day in the midst of winter. The increased photoperiod resulted in a complete remission of depression symptoms in 30 of 34 patients within 2 to 4 days of treatment. By selecting the exact time for bright light exposure, the researchers were able to manipulate the circadian rhythm of melatonin.

Although this approach may appear quite unusual, manipulation of light-dark cycles has long been an established method of influencing growth cycles in plants and activity patterns in animals—for example, in production of chicken eggs. Plants determine the time of year by measuring the length of daylight. In animals, a single exposure to bright light during the inactive phase of the day, if interpreted by the biological clock (brain) as a dawn signal (i.e., a new day is starting) can initiate a period of activity. In humans a neural pathway from the retina to the hypothalamus, where the biological clock is situated, has been found, suggesting that response to light exposure may be a purely physiological effect. Bright light at nighttime can cause insomnia in day workers but can help night workers synchronize to their schedules. Artificial light might even be helpful in dealing with jet lag. As researchers learn more about the effect of these cycles on humans, sleep clinicians may start taking a light-dark history from their patients to determine why they have been deprived of these synchronizing agents.

## Circadian Rhythms in Clinical Medicine

Almost all physiological variables display circadian rhythms—a waxing and waning apparent only if frequent measurements are taken across the 24-hour day. Thus it is apparent that one measurement of body temperature or one blood sample may not be sufficient to determine whether a value is abnormal. For example, urinary electrolyte excretion is five times greater during the daytime than during the night, while growth hormone shows an opposite pattern: It is rarely secreted in the daytime and almost always peaks in the first hours of sleep. Some chronobiologists have developed chronograms—plots of the range of normal values of a physiological parameter for each of the 24 hours. This information could potentially enable a physician to take one body temperature measure or blood sample, check the time on his watch, and consult the chronogram to see if the temperature or blood value is normal for that time of day. The difficulty with this approach is that it assumes that all patients share the same circadian physiology. Not only are there innate differences among individuals, but a patient's sleep-wake cycle, as determined by his work schedule, may have altered his circadian patterns before the doctor takes his measurement.

In some areas of medicine, careful consideration is given to circadian rhythms. Diagnosis of Addison's disease (decreased adrenal secretion) and Cushing's disease (excessive adrenal secretion) routinely takes into account the time of day when plasma cortisol samples are taken. Cortisol is secreted episodically, with highest values in the early morning close to awakening and lowest values in the evening near bedtime. This assumes that the patient is maintaining a normal sleep-wake schedule. In general, however, physicians have not yet taken into account the potential of the new information about circadian principles. This may not be surprising since their own disruptive on-call schedules as residents and interns are steeped in tradition and remain unchanged despite the availability of sound chronobiological alternatives.

There appears to be a marked circadian rhythm in human susceptibility and resistance to allergic substances, such as dust and pollen, and to chemical irritants, as well as in response to toxins and medication. The effect of alcohol is also influenced by biological time. Diurnally active healthy persons rated themselves as being more strongly affected by alcohol intake at 11:00 P.M. than at 7:00 A.M. or 11:00 A.M. Their reports corresponded with the results of performance tests in which the effects of alcohol on the nervous system were measured. Given the same dose, performance was more impaired in the evening.

Animal studies suggest that humans may show a marked variation in sensitivity to chemicals and x-rays used in cancer therapy. For example, *cytarabine*, a drug used in treating leukemia, was given to mice during their active (wake) and quiet (sleep) phases. With the same dose, 74 percent of the mice died at one phase, while only 15 percent died when the drug was administered at another phase. Similarly, mice subjected to the same x-ray dose at different times of day had different mortality rates: All mice exposed during their sleep period died, while all those exposed during their active period remained alive. Clearly, sensitivity to treatment varies markedly depending upon *when* treatment is administered.

One of the major problems in cancer treatment is that chemotherapy affects not only the malignant cells, but the healthy host cells as well. In humans there may be an optimal time for cancer chemotherapy and radiation based on the activity cycle of the tumor and host cells. Chronobiologists would try to administer the maximum treatment dose at the peak activity time of tumor cells and at a time of least toxic effect on host cells. This would give the maximum therapeutic effect and the fewest side effects. In addition to cancer agents, actions of a variety of other drugs—antihistamines, anesthesias, analgesics, aspirin, steroids—all show that differences in the time of administration make for significant differences in effect.

### Chronobiology of the Life Cycle

The timing of birth and death may be related to biological rhythms. Labor most often begins, for example, and natural births most often occur, between midnight and 6:00 A.M. This peak may reflect a circadian rhythm of various hormones. Induced births, in contrast, peak between 10:30 A.M. and 5:30 P.M., presumably reflecting a social rhythm—the preference of obstetricians for a daytime work schedule. Stillbirths and births that involve complications for the mother also peak during daylight hours. These data suggest that hospitals should encourage nocturnal deliveries and make sure that alert nursing and medical staff are on duty at these hours.

There is also evidence of a rhythmic pattern to the timing of death. The peak time of human deaths is around 6:00 A.M.—one of the likeliest times for REM sleep (see Chapters 5 and 6)—with a secondary peak at 4:00 P.M. It is as yet unclear whether these patterns reflect an underlying physiological rhythm or are merely determined by social factors such as availability of alert medical care at certain hours. A rhythmic phenomenon is identifiable as a biological rhythm only if it continues to show cyclic behavior in the absence of social cues.

Cycles longer than the circadian (daily) rhythms we have been discussing are much more difficult to measure and are also more difficult to link to the internal biological clock. Examples of circannual rhythms that may be biologically based include the monthly distribution of births, which has been carefully studied by the World Health Organization. The peak birth period in the northern hemisphere comes during the last week of July. Whether this reflects a natural fertility rhythm is unclear. Populations not using birth control show different peak birth months from those that do. Practices of sexual abstinence during certain religious holidays may also play a role. More study may yield the answers.

Circannual rhythms in deaths from certain diseases have also been documented. Mortality from cardiovascular diseases is highest in the winter months—January in the Northern Hemisphere and June in the Southern Hemisphere. Deaths due to respiratory diseases—asthma, bronchitis, influenza, pneumonia—peak in February in the Northern Hemisphere and in July in the Southern Hemisphere. These cycles may be linked to cold temperatures—the obvious connection—but can also reflect an underlying biological cycle that fluctuates on a yearly basis.

Just as the common wisdom that sunlight caused the leaves of de Mairan's plants to open was wrong, so too may our concept that cold environments cause disease be wrong. We may be more governed by our rhythms than we realize.

# 3

# Shiftwork:
# Clocks in Collision

*Whether you work by the piece or work by the day, decreasing the hours increases the pay.*

—Slogan of Advocates of Better Shiftwork Schedules, circa 1890

Human beings are a diurnal species—active by day and asleep by night. Yet over twenty million Americans struggle with work schedules that include night work—and they really struggle. Nuclear power and chemical plant operators, Strategic Air Command specialists, doctors, nurses, factory workers, police—all must operate around the clock.

Shiftwork is as old as recorded history. Since ancient times there has been a need for some members of society to remain awake throughout the night to provide security and to keep watch over property. This has never been an easy task. It is clear that the human species has evolved with the need for a daily sleep period of approximately one-third of the 24-hour day and that the preferred time is at night. Do night workers tend to be sleepy—and do sleepy workers pose a safety risk? The evidence is strong: The Three Mile Island accident, the Union Carbide explosion in Bhopal, and the Chernobyl disaster all occurred during the late evening/night shift.

Not surprisingly, I have learned from talking with them that most rotating and night-shift workers are unhappy with their schedules. It's midnight. I enter the plant cafeteria. Fifty shiftworkers sit there waiting for me to start my presentation. They're wondering what some doctor from Stanford could possibly know about shiftwork. I am wondering how they will receive me. After a brief introduction by the plant manager, I begin: "Who designed the shiftwork schedule at your plant?" The answer may be "someone who didn't work shiftwork," or "management," or "the secretary," and the list goes on. Actually, in most cases, the shiftwork schedules have not been designed by anyone in the company. They are based upon tradition handed down from generation to generation or copied from a nearby plant.

In a 24-hour facility, employees' lives and management policies revolve around the work schedule. Shiftworkers arrange their babysitting, carpooling, shopping, vacations, social events, and recreational activities around their schedules. Managers structure their pay, staffing, overtime, and operational policies around the schedules. Employees and managers do everything they can to adjust to their current schedule, even though it may be the wrong one for that particular industry. Because of the compromises and unquestioning acceptance of traditional schedules, it is difficult for either group to focus on the core problem—the schedule that they have all inherited.

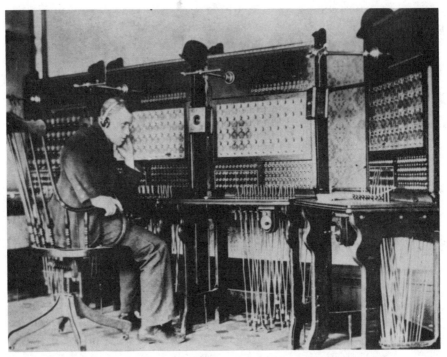

The management of the telephone company in the 1880s thought it unsafe to employ women in the office at night, so all night operators were men.

## 1883: A Big Year for Shiftwork

Eighteen eighty-three was a milestone in the history of shiftwork. Edison's invention of the light bulb and subsequent lighting up of one square block in New York City created a constant source of illumination and electricity and dramatized the new possibilities. For the first time it was possible for large numbers of people to work around the clock. This proved a major jolt. Not only were critical services such as police and fire departments expected to operate 24 hours a day, but any industry that could increase the return on its investment by increasing productivity was expected to double its working hours. Eventually, consumers looked for an ever greater number of services to be available at all hours of the day or night. Instead of being a diurnal species, many humans were moved by society into a new time dimension. Not the rising and the setting of the sun but work schedules determined when they were active and when they slept.

Edison's genius did more than provoke a major social change. It also upset nature's carefully conceived biological clock. The hundred years since his invention of the light bulb represent a mere instant in the evolutionary time scale. Over the millennia the human biological clock developed in harmony with the natural environment to synchronize with the subtle changes in day length—from ten hours in winter to fourteen hours in summer. Instead of synchronizing our internal clocks with these natural rhythms, suddenly our biology was asked to cope with the abrupt time changes caused by shiftwork.

Until round-the-clock light became available, most large manufacturing plants had only one team of employees; they averaged an 11-hour workday during the winter and, as daylight hours expanded, up to a 14-hour day during the summer. The capacity of the biological clock to adapt to new hours enhanced adjustment of internal rhythms to changing work schedules and day lengths. During the summer months the internal biological clock would gradually adjust to wake employees up at 5:00 A.M., while in the winter months it could program them to awaken at 6:00 A.M. to 7:00 A.M. It did not have the capacity, however, to handle the much wider swings caused by 24-hour work schedules.

With the development of new technologies that demanded continuous operations to avoid expensive shut-down and start-up procedures, 24-hour coverage became widespread. The capital-intensive and "never-put-out-the-fire" industries, such as iron foundries and steel mills, were among the first to rely on round-the-clock operations, 24 hours per day, 7 days per week. In the steel mills of the United States—one of the largest continuous industries—the most common shift schedule was for employees to work 12 hours a day, 7 days a week. This 84-hour week made it possible to have 2 employees covering one operating position for the full 168 hours per week. One employee worked from 6:00 A.M. to 6:00 P.M., while his co-worker covered from 6:00 P.M. to 6:00 A.M. Employees rotated between the day and night shifts every two weeks. To accomplish this changeover, one employee had to work 24 consecutive hours at the end of every 28-day cycle, while his co-worker got those 24 hours off. This happened on Sunday and represented the only time off each month!

If two employees, Mr. Smith and Mr. Jones, covered one continuous job, the shift schedule for one would look like Figure 3-1.

FIGURE 3-1

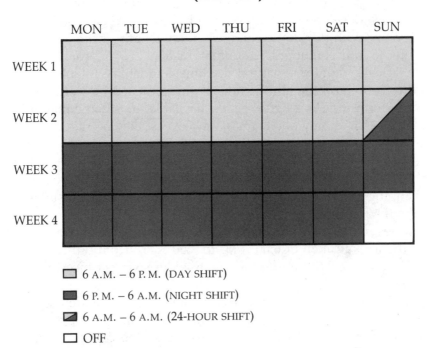

**12-Hour-Day, 84-Hour-Week Schedule (1883–1920)**

□ 6 A.M. – 6 P.M. (DAY SHIFT)

■ 6 P.M. – 6 A.M. (NIGHT SHIFT)

◪ 6 A.M. – 6 A.M. (24-HOUR SHIFT)

□ OFF

Although 100 years later shiftworkers have more time off, this 28-day cycle is still the foundation for most shiftwork schedules in U.S. industry today.

It is hard for present-day American shiftworkers to grasp the fact that the shiftworkers themselves developed such schedules. Many were European immigrants who not only put up with the arduous schedules but favored them because their wages—about 15 cents per hour—were much higher than any they had been paid in their home countries.

Eventually these difficult work schedules contributed to the rise of unionism and the Homestead (a Carnegie Steel Company plant) strike of 1892 by the Amalgamated Association of Iron and Steel Workers of America. This union was one of the largest in the United States at that time, representing 24,000 of the 400,000 total union members in the country. Fighting between Pinkerton guards and striking workers killed thirteen workers, thrusting labor problems upon the consciousness of the federal government. Many of the specific strike issues centered around the workers' demand that the Carnegie Steel Company recognize the union. Charles B. Spahr reported in *America's Working People*, "It was not the lowering of wages that caused the most bitter complaints among the men. Their wages, even when lowered, were not low and most of them realized it. Their real grievances were the long hours, the Sunday labor, the strain under which they were compelled to work. . . ."[2]

One of the ironies of the Homestead strike was that workers in two of the Carnegie Steel Company's older plants (the Homestead plant had been a competitor until it was purchased in 1889) had much better schedules. In 1873 Andrew Carnegie and his partners had imported the Bessemer steel process from London to Pittsburgh and hired Captain William R. Jones, a gifted steel master, as the plant superintendent of the Thompson Works. ("Captain Bill" was paid $25,000 a year, the same salary that the president of the United States earned.) He scheduled his men on 8-hour shifts; this, together with his knowledge of the new steel technology, accomplished outstanding improvements in productivity. Captain Bill reported that it was "entirely out of the question to expect human flesh and blood to labor incessantly for twelve hours, and therefore it was decided to put on three turns, reducing the hours of labor to eight."[3]

**Workers at the Bethlehem Steel Corporation tended a "never-put-out-the-fire" operation.**

This was a remarkable discovery. At that time, 12-hour workdays were the norm. Most Americans were farmers and accustomed to working long hours. Because no one had anticipated that industrial work around the clock might be more fatiguing than farm work, steel mill managers were surprised to find that productivity actually increased when employees worked fewer hours. In spite of this discovery, the 12-hour day, 7-day week not only continued but expanded in some industries during the early 1900s. In 1907, 97 percent of all employees in blast furnaces were working this schedule, and during the industrial boom of World War I the number of manufacturing shiftworkers on the schedule doubled.

The greatest pressure for change came not from the shiftworkers but from religious and civic leaders. In July 1920, the Interchurch World Movement released a report on the steel industry: "Americanism is a farce, night schools are worthless, Carnegie libraries on the hill-tops are a jest, churches and welfare institutions are ironic, while the steel worker is held to the 12-hour day. . . . Not only has he no evening left, he literally has no time left after working such schedules. . . . Americanization of the steel worker cannot take place while the 12-hour day persists." The Interchurch World Movement report concluded that the continuous 12-hour shift contributed to high rates of industrial accidents and poor health to such a degree that most shiftworkers were dying before the age of fifty.[4]

A magazine reporter of the time described the shiftworkers in the following way: "They came out in groups from the mill gate at the change of shifts, fagged after their twelve hours before the furnaces, Turkish towels about the open necks of their shirts, wet in the back to the belt. At every corner, the group would grow smaller as the men turned into their homes to wash up, eat, go to bed, get into their mill clothes, and go in on their turn again."[5]

Justice Louis D. Brandeis of the U.S. Supreme Court perhaps put it most eloquently in testifying on the 12-hour shift before a committee of the U.S. House of Representatives in 1912:

> It has been stated, and there seems to be much evidence to support it, that such a life as this makes a man old at 40. The astonishing thing is that they should live until 40. I think they must be of peculiarly hardy build and of excellent constitution to be able to stand such a life as they do and to work until 40 years.
>
> But it is not merely the dependence of that individual, it is not merely the fact that he becomes a useless individual and a burden to his family at 40, it is the fact that he is the father of a family and transmitting through another generation, and perhaps through many generations, the evil weaknesses and the degeneration which have come to him

through the life to which he has been subjected—and it is not merely that; the degeneration is not merely physical; it is a degeneration which is moral as well, and how could it be otherwise! I ask you, gentlemen, to remember that these persons, however they may differ from us in race or in their habits of living, are individuals. Imagine what would be our condition if we, 7 days a week, undertook to work 12 hours a day.

I submit there is only one thing in which we could find relief, and that is in some form of dissipation. We could resort to it, and when you look into the lives of these men you will find how nearly allied they are to us, because they resort to precisely that thing.

Mr. Fitch has called attention to the fact of the use of liquor and to the expenditures of these men for liquor after pay day. In a single city it appears that these liquor dealers have gathered together at pay day from the hard-earned wages of these people from $30,000 to $60,000, which amount is deposited in banks by the liquor dealers after the lapse of a day or two from the paying off of the men.

What does that mean? It does not mean merely the waste of that money, but it means there is going on a moral degradation of those persons and of persons connected with them. It is not true, Mr. Chairman, that the burden is borne wholly by their families—fortunately it is not. It is borne by the community.[6]

### A New Era

In 1914 Henry Ford announced a daily wage of five dollars for an 8-hour day on his production line. Instead of his going bankrupt as many predicted, the resulting productivity records—a new car every twenty seconds—underscored for the rest of industry the inescapable truth of the opposite side of the coin: The ill effects of poor scheduling were reflected in lower productivity and profits. The modern age of shiftwork scheduling was ushered in, and the 8-hour work day became a standard in many industries. But some of the older continuous operations, bound by tradition, remained under the domination of the 84-hour work week.

In 1919, shortly after the end of World War I, steel workers and miners went on strike for union recognition and reduction of the 12-

hour working day. Secretary of Commerce Herbert Hoover convinced President Harding to step in. With support from civic and church groups, Harding made a series of speeches against the 84-hour (12-hour, 7-day) shift schedule. The president was able to obtain assurances from industry executives that they would phase out the old work schedules. Ironically, many employees, fearful of a drop in income, wanted to keep the 84-hour week schedule. As Andrew Carnegie saw it, "The 12-hour day was established by the men long before the United States Steel Corporation was organized."

On August 3, 1923, the iron and steel industries adopted plans to eliminate the 84-hour week. Shiftwork operations were divided into 3 teams, each working 8 hours a day, 7 days a week (56 hours per week). To implement the new schedule, a 25 percent hourly wage hike was effected, so that while the work week was reduced by a third, weekly wages were reduced by less than 17 percent. As a result, the average work day for shiftworkers in the United States dropped from 72.3 to 59.8 hours per week in the period from 1922 to 1926.

In the first forty years after the invention of the light bulb, shiftwork schedules had their greatest visibility. Large numbers of Americans had for the first time been asked to work long hours around the clock. The high rate of industrial accidents and deaths, reduced life expectancy, and decrease in performance made it apparent that there are limitations to the capabilities of the human organism to operate day and night.

Although the reduction of the work week to 56 hours was an important improvement, it is not surprising that shiftwork schedules failed to take into account the limitations of sleep-wake cycles. Little research knowledge existed. If an employee was scheduled to work from midnight to 8:00 A.M., it was assumed that he would be able to do so and to operate efficiently. During the Depression, with the passage of the Walsh-Healey Act, the standard work week was reduced to 40 hours. Shiftwork schedules were changed again, so that now 4 employees, each averaging about 42 hours a week, covered one continuous job. Since there are 168 hours in a week, a 42-hour work week makes sense for continuous operations ($42 \times 4 = 168$). However, most work policies are based on day-worker concepts: Monday–Friday, 9:00 A.M.–5:00 P.M. ($8 \times 5 = 40$); hence the 40-hour work week. As a result, shiftworkers are paid overtime premium pay for the extra hours.

# 12-HOUR DAY ENDS FOR STEEL INDUSTRY

## All Members of Institute Agree to Make Change Without Delay.

## WAGE SCALE IS ARRANGED

### Workers to Get Higher Pay Rate, but Will Earn Less—Costs to Go Up 15 Per Cent.

Plans to abolish the twelve-hour day in the steel industry, adopted at a meeting of the directors of the American Iron and Steel Institute at the Metropolitan Club yesterday, will be pushed forward without unnecessary delay, said Judge Elbert H. Gary, President of the Institute, in a statement following the meeting.

The action taken by the Institute was the clumination of a series of conferences which began late in June after President Harding had asked the Institue to pledge that it would abolish the twelve-hour day when conditions of labor warranted that course. The President's request was made on June 18 and on June 27 the directors of the Institute wrote to the President saying they were determined to exert every effort at their command " to secure in the iron and steel industry of this country a total abolition of the twelve-hour day at the earliest time practicable."

During World War II many individuals were thrust into shift-work, but again little attention was paid to the schedules themselves except by a small group of social workers and industrial engineers who developed an interest in shiftwork as a legitimate field of scientific inquiry. Studies of the effects of shiftwork during this era revealed a higher prevalence of ulcers and sleep disorders in shift employees than in day workers. Social and family problems were also common.

Shiftwork continued to increase steadily after World War II in the United States and in other industrial countries. As of 1982, in the United States, 26 percent of adult males and 16 percent of adult females were employed at jobs that required them to shift between day and night work. Inevitably, sleep disorders afflicted many of these workers, but at that time there were few sleep clinicians to help them; the field was virtually nonexistent.

With the discovery in 1953 of active physiological sleep stages (see Chapter 5), sleep research grew rapidly and soon gave rise to the clinical discipline of sleep disorders medicine. In 1969 Stanford University opened the first sleep disorders clinic in the United States. For the first time doctors opened their doors to treat patients with a wide array of complaints, not only of poor sleep but also of impaired alertness. Opening the first clinic of a new medical specialty is a risky undertaking. No one was sure what types of patients would come or what cures were available. Many patients came with unusual complaints, often representing the first such case studied. Among this group were several shiftworkers complaining of an inability to adjust to their work schedules.

One man who operated a forklift truck rotated nights, days, and evenings. In order to stay awake during his 8-hour shift he would drink ten to fifteen cups of coffee. At home he would frequently fall asleep while watching TV and reading the newspaper. His wife finally dragged him into the sleep clinic after he fell asleep while driving home from work. Our treatment approach for such people would be to look for an underlying sleep disorder such as narcolepsy or insomnia and to treat these disorders with medications and counseling. When such a disorder was not apparent we might write a letter recommending that this person be reassigned to day work because of his inability to adjust to shiftwork.

Our approach was straightforward but limited: Shiftwork causes sleep and alertness problems; the solution is not to work shifts. This did not, however, alter the fact that someone would have to work nights—I don't think we realized how many. Although the link between sleep disorders and shiftwork was established, there was no real understanding of the magnitude or cause of the problem. Academic research on shiftwork was also limited, consisting largely of scores of papers indicating that shiftworkers had more sleep problems than day workers.

This observation was confirmed in 1980 by a phone call from Preston Richey, operating manager of the Great Salt Lake Minerals & Chemicals Corporation in Ogden, Utah. Mr. Richey called the Stanford Sleep Clinic after seeing a newspaper article on a sleep study that said "these findings may have implications for shiftwork"—a common conclusion of sleep research articles, but one that rarely leads to change. Mr. Richey said he had 150 employees who were having trouble sleeping. He had looked in the local library and the university library, and had called and written to the U.S. Bureau of Labor, attempting to discover the best way to handle shiftwork. What he found was lots of information on the problems of shiftwork but few answers.

## The Great Salt Lake Study

The Great Salt Lake Minerals & Chemicals Corporation (GSL) is located on the edge of the Great Salt Lake. The plant harvests potash from its evaporation ponds in the lake and processes it at its plant. At the end of 1980, Mr. Richey held individual meetings with each of his employees to identify their concerns about their jobs. To the surprise of management, the number one complaint raised was shiftwork.

A few months later Dr. Charles Czeisler and I left the Stanford Sleep Clinic and flew to Utah to meet with the management of GSL. It was an interesting meeting: two sleep specialists from Stanford talking to a plant manager, president, and vice-president of a chemical plant about sleep research and shiftwork. We learned Mr. Richey's primary reason for calling upon expert sleep researchers: to assure the employees that the company's shiftwork schedule was the best that could be devised. "We really didn't want to change the schedule; we wanted employees to accept it. I didn't think we could do any better."

We saw our task differently: It was to convince management that the available research knowledge was sufficient to enable us to improve and redesign the company's shiftwork schedule. Although several managers initially did not want to change schedules, we garnered enough support to start up a study. In fact, one of the vice-presidents later said he hired us because of our enthusiasm rather than for any specific new schedule ideas (which we didn't yet have!).

Our first step was to analyze the current shift schedule. For 9 months of the year, the plant operated 24 hours a day, 7 days a week. During these months each employee rotated shifts from day (8:00 A.M. to 4:00 P.M.), to night (midnight to 8:00 A.M.), to evening (4:00 P.M. to midnight). During the summer months, when all the potash had been harvested, the employees worked only a straight day shift. This may have contributed to their intense dissatisfaction with shiftwork: When they went back on a shift schedule, they remembered how well they had felt on a normal day schedule. On the other hand, many shiftworkers who rotate all year long forget what it feels like to be well rested and alert.

The shift schedule operated in the following manner: Each employee worked seven night shifts in a row, followed by two days off;

## FIGURE 3-2

### 8-Hour-Day, Weekly Phase Advance Schedule

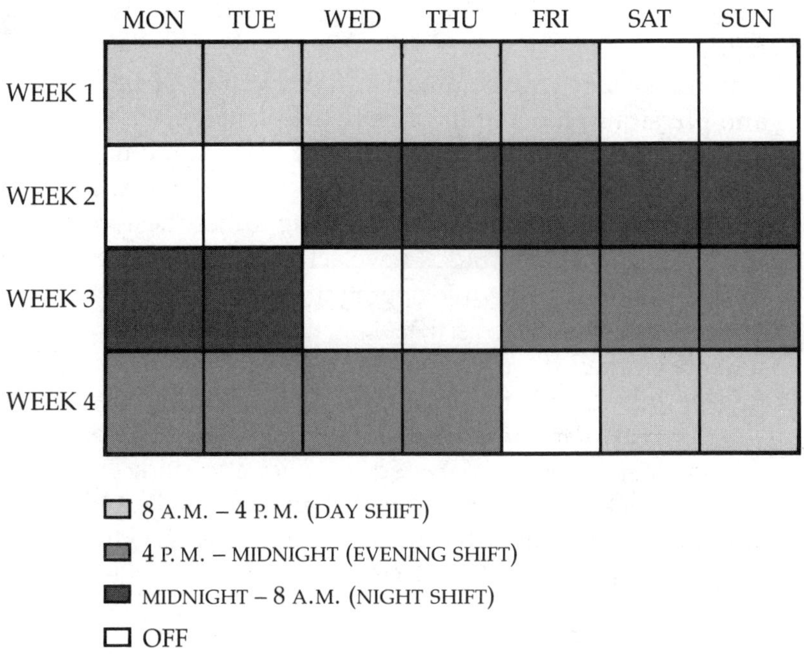

8 A.M. – 4 P.M. (DAY SHIFT)

4 P.M. – MIDNIGHT (EVENING SHIFT)

MIDNIGHT – 8 A.M. (NIGHT SHIFT)

OFF

seven evening shifts in a row, followed by one day off; then seven day shifts in a row, followed by four days off. Each shift lasted eight hours. The four-day respite—Saturday, Sunday, Monday, and Tuesday—was called the "long change" and was the only weekend off in four weeks. We called this schedule a "weekly phase advance," because it forced employees to advance their work schedule by eight hours in the counterclockwise direction each week. It included a stretch in which employees worked 14 out of 15 days, and the average work week was 42 hours. We later learned that this was the most common shiftwork schedule in the United States (see Figure 3-2).

Gordon Thompson, shiftworker on a standard counterclockwise rotation, in testifying before the U.S. House of Representatives on March 23, 1983, described the difficulties inherent in the weekly phase advance schedule.

> When I was employed in 1973 by Firestone Tire & Rubber Company, the only shift I could obtain at that time was a seven day rotating shift. The shift consisted of working seven nights (12:00 P.M. to 8:00 A.M.), off two days; working seven evenings (4:00 P.M. to 12:00 P.M.), off one day; working seven days (8:00 A.M. to 4:00 P.M.), off four days. I worked this rotating shift for approximately two and one-half years.
>
> It did not take long before it started affecting me mentally and physically. Most of the jobs in the plant are very physically demanding jobs; therefore, rest became the number one priority. At first I had problems sleeping and the rest that I got did very little good. Working under these conditions was very dangerous. I have actually seen workers go to sleep standing up while operating a piece of equipment. Even driving home was a task within itself, particularly when I worked the midnight shift; several times I caught myself dozing and running off the side of the highway.
>
> After two years of working this rotating shift, I started taking sleeping pills. It started to affect my eating habits; my sense of taste gradually diminished. I also had problems with my stomach and digestive system. It was very difficult to decide what type of food to eat, especially on the 12:00 P.M. to 8:00 A.M. shift.

After a year of rotating, I started having family problems because I had hardly any time off to be with them. Before I started rotating work, I used to hunt, fish, and enjoy sport activities; that all stopped. I lost interest in everything that I had previously enjoyed. Not only did it affect me, but everyone I worked with.

I could see such a difference in my friends at work. People would go to the break area and no one would talk to each other; and if they did, it was something negative about their work. The working relationship with the supervision was also very poor. It seemed like it was a constant battle between production workers and the management.

It seemed as though every week I would hear of someone getting a divorce or going to get one. I thought I would get accustomed to working under the conditions, but the longer I stayed, the worse it became. After three years, the employees' attitude, absenteeism, work-related injuries and low productivity became so bad, the company decided to abolish the seven-day rotating shift. It was like working in a different plant.

From my own personal experience with all the stress and physical and mental abuse, if I had to go back to the seven-day rotating shift tomorrow, tomorrow would be my last day to work.[7]

Most companies using this schedule are unaware of its origin. A manufacturing company in Pittsburgh that I recently visited discovered that all their plants in the United States were on this kind of schedule. None of them could recall actually choosing the schedule; they surmised that it had originated in the early 1900s when the company was first organized. This particular schedule is called "The Southern Swing," "The Hoover Schedule," "The Dow Schedule," "The Hanford Schedule," and a variety of other names, depending upon the local tradition. At one refinery working this schedule, the plant manager told me that the Hanford schedule was developed by a Mr. Hanford, while the operating manager confided that it was named after Hanford, Washington, where it was first used.

It became clear to us that the schedule being used at GSL was based on tradition, with many of its features derived from the original schedule of the 1890s. It was, in fact, the traditional schedule divided in half. Instead of two workers covering a position, there were now four. Each employee averaged a 42-hour week (thus allowing for coverage of 168 hours in each week) instead of 84 hours. Instead of rotating shifts every two weeks, workers now rotated shifts every week. The GSL schedule took 28 days to cycle, just as the old steel mill schedule did.

Although the weekly phase-advance (counterclockwise) schedule adequately met the company's need for continuous operation, there was little emphasis on employee preference and no consideration of those physiological principles that we believed could influence adjustment to shiftwork. The symptoms caused by this schedule were similar to those reported by air travelers when traveling from California to England. Unfortunately, the shiftworkers experiencing this "blue-collar jet lag" did not have the benefit of seeing London—and their bodies were scheduled to make the arduous trip every week!

One of our own key circadian principles is that shift schedules should rotate to later hours, in a clockwise fashion. For example, a worker who goes to sleep at 10:00 P.M. while working day shift will find it

## FIGURE 3-3

### 3-Week Rotation in the Right Direction

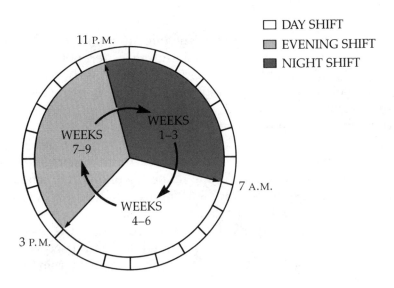

☐ DAY SHIFT
▨ EVENING SHIFT
■ NIGHT SHIFT

11 P.M.

WEEKS
1–3

WEEKS
7–9

7 A.M.

WEEKS
4–6

3 P.M.

easy to delay bedtime by 4 hours, until 2:00 A.M., when moving to the 4:00 P.M. to midnight (evening) shift. This is a clockwise rotation. But going from evening shift to day shift is a counterclockwise rotation, which forces employees to move their bedtimes to earlier hours. This is equivalent to creating a 21-hour day, requiring an adjustment beyond the capability of the 25-hour clock. To change your bedtime to 10:00 P.M. after getting used to falling asleep at 2:00 A.M. is not easy. Imagine a straight day worker, who normally goes to bed at 11:00 P.M., trying to get into bed four hours early—at 7:00 P.M.

Our research team decided to implement two schedules: One group of employees was put on a weekly phase-delay (clockwise) schedule. Instead of the old method of rotation, they now moved to later hours—from days to evenings to nights, and back to days. This clockwise rotation was selected because it was compatible with the free-running rhythm of the biological clock. Since we naturally drift to later hours it makes sense that the shift schedule should rotate to later hours. The new schedule actually forced employees to delay their bedtimes, but it still had the disadvantage of forcing an abrupt change every week.

A second group of employees was scheduled not only to change their direction of rotation to later hours, but also to decrease the rate of rotation. Instead of rotating every week, these employees changed shifts every three weeks (see Figure 3-3). Furthermore, to avoid the abrupt 8-hour shift change, we instituted a "slow-drift rotation." For about three days during shift transitions, each employee was scheduled to come to work 2½ hours later than on the previous day. By means of this slow-drift rotation they could make a gradual transition to their new shift. Forced to free-run between shift transitions, they lived on a 26.7-hour day for three days—close to the natural free-running rhythm.

At first our recommendations appeared quite strange to GSL managers, as they were familiar only with their traditional schedule. Our schedule incorporated not only operational needs and employee preferences (we put in a new day-off cycle with more time off) but circadian principles as well. Employee support was strengthened by our meetings with each crew, which allowed us the opportunity to explain the application of these biological principles. Finally, we won enough management and employee support to initiate a three-month preliminary trial of the schedule.

The first problem we ran into was opposition to the slow-drift rotation. The employees disliked it, mainly because it was unusual, hard to get used to, and—most important—their families could no longer figure out when they were working and when they were off. Some families must learn to cope with these schedules. In Sweden, for example, there are police forces that traditionally work a schedule whereby each day they report to work four hours earlier than the previous day. We did, however, eliminate this slow-drift feature, and the employees reverted back to about one 8-hour abrupt shift change every three weeks.

The three-month trial was successful and was extended. Our research team went back to GSL several times to evaluate the impact of the schedule. We prepared a range of questionnaires, and the plant management selected several productivity measures to aid the evaluation, such as the number of tons of potash loaded and hauled away from their solar pond per hour, as well as the number of tons of potash produced in the plant per hour. We asked them to review their records for the two years before and the two years after our intervention. Since the data was not computerized, an intimidatingly huge stack of raw data was placed before us. Fortunately, a group of shiftworkers volunteered their time to sort through it, and within two days, with about eight people working around the clock, we were finished.

The results were surprising to everyone. Productivity had improved by 20 percent. Furthermore, when we compared the two schedules, we found that dramatic improvements in job satisfaction, schedule satisfaction, and health were apparent only in the employees on the three-week rotation cycle—those who rotated shifts every three weeks as opposed to every week. Those who had merely changed the direction of rotation showed less dramatic improvements.

The major finding from the study was that shiftworkers not only preferred schedules that were consistent with circadian principles, but also that they were more productive on such schedules. As these new concepts were implemented, however, several unforeseen problems arose. It became clear that in order to make the new schedules work, management might need to make some operational changes, such as reassigning supervisors, while employees might have to become more flexible in their attitudes toward job assignments.

We also learned that to implement change was much harder than we had anticipated. When a new schedule is first presented, the work force tends to seize upon any obvious negative features. The positive, built-in features—those that improve health, increase alertness, and just make you feel better—are not appreciated until the schedule is tried. Holding up a piece of paper on which the new schedule is printed does not show you how much better you'll feel. Education about shiftwork scheduling, sleep research, and circadian principles prior to the schedule implementation, however, persuaded the employees to give the new schedule a fair trial. Over a year's time, employee acceptance of the new schedule grew from 60 percent to 90 percent.

Typical of employee response to the new schedule was that expressed in the testimony of shiftworker Scott N. Nielsen at a U.S. Congressional committee hearing.

> This system [the new schedule], I feel, has worked out quite well. Of course there are always those who are against change just because it's change. But over the last two years, everyone seems to have accepted the idea.
>
> In the last two weeks I have been talking with my fellow workers trying to get some input to help me in preparing for this statement. I have found that everyone has said pretty much the same thing. They like it a lot better than the old schedule. Some said that it seemed like they had a lot more time to do the things that they wanted to do, instead of just "eat, sleep and go to work." I asked why. The response was that they have a chance to get used to a shift and can get by on a lot less sleep, which gives more free time. I know that the attitude improvements have been great. Morale has stayed consistently high. When morale is high production is high. Another thing that I have noticed since the schedule change is the amount of interdepartment bickering and backbiting has come to a complete stop.[8]

### Other Shiftwork Strategies

There are other strategies for coping with shiftwork. One approach is straight shifts, whereby each employee is given a permanent assignment to an 8-hour time block. The major physiological problem that accompanies this fixed schedule is that the night workers typically never adjust their biological clocks, because each week during their time off they revert back to a day schedule. The conflicting time cues—"be alert during the daytime while off work," "be alert at night while working"—make any type of physiological adjustment difficult.

From an operational point of view, straight-shift schedules can create training problems. Many employees would prefer to remain at a low-level job on day shift rather than accept a promotion that might assign them to the night shift. Often employees who dislike supervision tend to gravitate to the night shift, leading to crew imbalance. The evening shift (3:00 P.M. to 11:00 P.M.) is generally the most disliked because it is most disruptive to family life. A mother on this schedule might not see her children except when they are asleep. The problem is compounded because the younger employees, who have the least seniority and therefore typically end up on this straight shift, are most likely to have school-age children. Despite its imperfections, a properly developed rotating schedule can be physiologically superior to a straight-shift schedule.

## FIGURE 3-4

### 12-Hour-Day, Average 42-Hour-Week Schedule (1955–present)

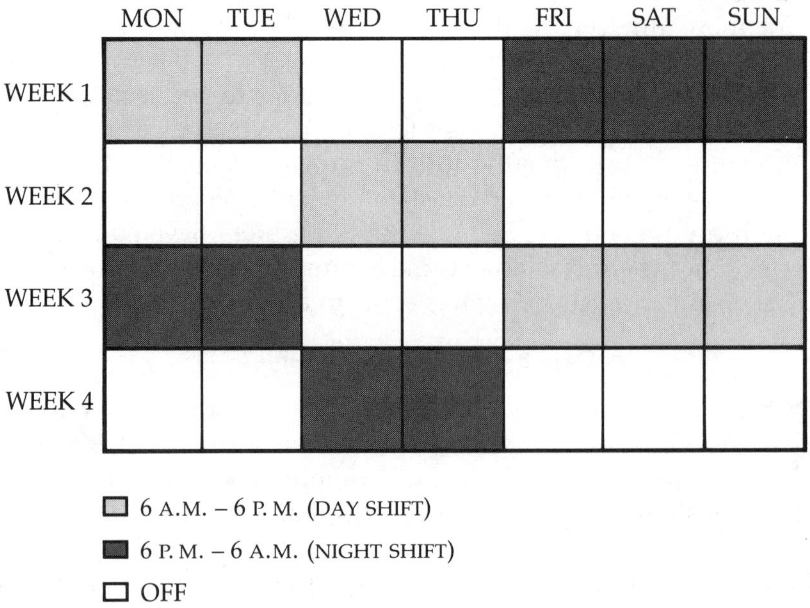

☐ 6 A.M. – 6 P.M. (DAY SHIFT)

■ 6 P.M. – 6 A.M. (NIGHT SHIFT)

☐ OFF

An approach to shiftwork adjustment widely used in Europe is rapid rotation: An employee may work one day shift, then one evening, followed by one night—all within three days. The logic behind this technique is that this rapid rotation allows circadian rhythms to remain synchronized with the day shift; employees do not spend enough time on the evening or night shifts for their internal rhythms to adjust to the late hours. The assumption is that one can never completely adjust to night work and therefore might as well tough it out and rotate quickly through the night shift. While it may be helpful for the employee's adaptation to a normal social life, this approach is unlikely to provide significant improvement in alertness on the job during late evening and night shifts. Our approach, on the other hand, attempts to provide adjustment to all shifts by proper scheduling and educational techniques.

### The 12-Hour Shift Comes Back

In 1955 the 12-hour shift returned to American industry. This time, instead of only two employees covering the position, four employees were assigned to each job. The Eli Lilly Company instituted this schedule at its plant when it increased its operations from 102 to 168 hours per week. A 12-hour shift schedule looks like Figure 3-4. Because employees work 12 hours at a stretch, they need report to work only 14 days out of 28. This results in much more time off and reduces commuting expenses, but there are drawbacks.

The problem with 12-hour shifts is that they are weighted heavily to employees' preferences with little regard to circadian principles. To work 12 hours in a row is hard enough, but the difficulties are compounded when the ninth, tenth, eleventh, and twelfth work hours occur between 2:00 A.M. and 6:00 A.M., the lowest point of the biological performance-alertness cycle. In general, 12-hour schedules are most appropriate in operations that are highly automated, where the alertness of an individual operator is either unimportant or is maintained by a sophisticated alarm signal that will awaken a sleepy employee. Industries such as mining or air traffic control, which require heavy manual material handling or sustained alertness, have generally stayed away from 12-hour shifts. By 1984 at least 60,000 persons were reported to be working 12-hour rotating schedules, primarily in the chemical and petrochemical industries. Hospital nursing staffs are also increasingly turning to 12-hour shifts.

Because it is beyond the capacity of the human biological clock to make rapid 12-hour shift adjustments (it takes anywhere from 4 to 14 days for the body temperature cycle to make a 180-degree inversion), employees do not attempt to adjust to the scheduled night shifts. Instead they try to tough it out. Not suprisingly, it is the younger employees who typically favor 12-hour shifts. They are more willing to sustain the lack of physiological adjustment—poor sleep, a significant drop in alertness—in exchange for more time off. The older employees, on the other hand, because they feel the effects of circadian maladjustment to a greater extent, find the trade-off increasingly unattractive.

Over the past few years I have evaluated shiftwork schedules in chemical plants, refineries, paper mills, manufacturing plants, and utility companies. Although the most common schedule is the weekly phase advance, I have found a wide range of unusual schedules. One of the most interesting was in a coal-generated power plant in Texas that operated 168 hours a week. Each employee worked ten days in a row, followed by a four-day break. The shifts rotated from days to evenings to nights—the clockwise direction—consistent with the biological clock. Because the ten days on, four days off schedule synchronized with a seven-day week, each employee had his own day-off schedule that repeated year after year. For example, those operators with the greatest seniority always had Friday, Saturday, Sunday, and Monday off, while a junior operator might get stuck with Monday through Thursday off. A few employees had not had a weekend off for many years. One complained that he never saw his son pitch in the Little League (Saturday was game day), where he had been a star player. Still the employees preferred this schedule because eventually, with enough seniority, they would move into the favorable day-off schedule. Additionally, they preferred a long break in which they could go hunting or fishing, rather than two days off—which they often spent in town doing chores at home.

This schedule was a good fit for this geographically isolated, rural facility, and, of all the 24-hour plants in the United States that I have thus far studied, this plant's employees expressed the highest levels of job and schedule satisfaction. From a physiological viewpoint their schedule is optimal: Ten consecutive days on a steady shift is enough time for the biological clock to adjust; night shift starts up only once every six weeks; the stretch of night shifts is uninterrupted by days off, so that the internal clock is allowed to fully synchronize. Furthermore, four days off between shifts is ample time to make the 8-hour shift adjustment, because the internal clock can delay by 2 to 3 hours per day. Another important ingredient in the success of this schedule was adequate staffing. Most shiftwork schedules have built-in ineffiencies; when no one is sick or on vacation plant managers may need to invent special projects for excess employees. At other times of the year, excessive overtime is used to compensate for absenteeism. Over- and understaffing typically cause morale problems.

Not surprisingly, our recommendation was for the company to stay on this schedule. Our attempts to make the day-off schedule more equitable were not supported, but educational meetings were held with all the shiftworkers to teach them some of the chronobiological techniques of adjusting their sleep-wake hours to each shift. By gradually changing their sleep and wake times, employees were able to synchronize internal body time with new work time. They and their families were taught how to manipulate sleep-wake rhythms at each point in their specific work schedules. The employees were happy to find out that their schedule was sound and did not require any major changes. Management, too, was pleased to have the assurance that theirs was an excellent schedule. One manager said, "Now we won't have to listen to any more complaints about shiftwork for the next ten years."

Another interesting assignment revealed how little weight physiological principles sometimes carry. A paper mill in rural North Carolina called me in to try to solve a problem: Their employees were working too much. In one department of the mill, the work load constantly changes, so that the number of hours an employee works also changes. To keep the mill running 24 hours per day, seven days a week, for stretches of several months, employees might be asked to work overtime; in other months the department might be operating only five days a week—or might even shut down for weeks at a time. Over the years, only employees willing to work huge amounts of overtime were accepted into the department. These employees, nicknamed "overtime hogs" by their co-workers, were accustomed to 70 pay hours weekly—almost double their base salary. Many of us probably know some employees who accept all overtime opportunities, but imagine one hundred of these workaholics in a single department! Their shift schedule was a simple three-week cycle: seven night shifts, seven evening shifts, seven day shifts.

Each employee worked seven consecutive days on one shift and then rotated to the next earlier shift (counterclockwise rotation). No days off were scheduled. Physiologically, this was one of the most difficult schedules I had encountered—rotating weekly in the wrong direction without time off to enhance adjustment. Of course, the benefit of the schedule was the huge amount of overtime. As long as the department shut down for a week every few months employees had some time off.

When I was called in, the recession of 1982 was ending, and it was projected that the mill would be running continuously without shut-downs for the next twelve months. In fact, several employees had already worked a whole year with no days off, not even national holidays—365 consecutive work days! Management feared that the current schedule would lead to accidents and health problems if it were continued for a year. Meetings with the shift employees, how-ever, indicated that despite evidence of poor sleep and difficulty in adjusting to shift changes, they were extremely "satisfied" with their schedule. Actually, these employees had become prisoners of their work schedule. With their premium pay, many had purchased motor boats, motor homes, recreational vehicles, and so forth. In order to keep up on their payments, they needed to keep working 56 hours a week. Of course, the big question was, When did they get to enjoy these luxuries?

After much debate, the employees agreed to try out a phase-delay schedule that rotated clockwise every two weeks and included at least four days off per month. Employees did not like working two consecutive weeks on the evening shift—3:00 P.M. to 11:00 P.M.—because during the summer months outside temperatures were often above 100 degrees, and inside the mill the temperature and humidity were often greater. After the trial, although they accepted the new day-off schedule, they preferred to go back to their old method of weekly rotation.

### The Future of Shiftwork Scheduling

Since the publicity centering around the Great Salt Lake (GSL) study in 1980, American industry has become increasingly aware of the research on biological rhythms and its applications to alertness, performance, and health. Ironically, the primary impact of the GSL study has been to prompt plants to change only the direction of rotation—the schedule change that by itself did not produce significant gains at GSL (changing both the direction *and* the rate of rotation had the greatest impact). Some plants make the change because clockwise rotation just appears to be more logical, but most plants change over because management is aware of the inherent 25-hour biological rhythm. The typical plant manager and personnel manager in continuous industries have read or heard about biological clocks. Some of the larger and more innovative corporations are addressing the problem of shiftwork in greater detail, trying to make shiftwork schedules more attractive than the traditional Monday to Friday, 9:00 to 5:00 schedule.

Two opposing forces affect the future of shiftwork. On the one hand, improved technology, automation, and robotics may eliminate many traditional shiftwork jobs. Some industrial jobs that now require hands-on work may be run by computers, and shift operators may simply monitor computers. In one pulp mill I visited, a shiftworker is assigned 24 hours a day to watch a television monitor showing the pattern of wood chips burning! On the other hand, as society increasingly demands 24-hour services—in banking, stock market trading, security, communications, or even the local Seven-Eleven stores—shiftwork is likely to expand from manufacturing to service industries. It is likely that a large segment of our society will be required to work around the clock.

Another major problem for shiftworkers is time off. There is less of it and it comes at the wrong time. When a shiftworker is absent from work, whether because of vacation, sickness, jury duty, or personal needs, his shift is typically covered by assigning a co-worker to a double shift. In contrast, when a straight-day worker is absent, his spot remains vacant. As a result the average shiftworker in a seven-day-per-week operation often works about 400 hours more per year than the typical day worker—the equivalent of ten more 40-hour work weeks per year. It is ironic that these workers, already facing social and alertness problems, are also overworked.

By moving away from traditional scheduling practices, it is possible to give each employee more time off without bringing in extra personnel. Unfortunately, "tradition is the albatross around the neck of progress." The traditional methods used to sweeten shiftwork, such as shift-pay differentials or premium weekend pay, are not effective in dealing with the employee's need for quality time off, better health, and the maintenance of a family and social life.

No perfect schedule can be devised to solve all the problems of shiftwork. In any continuous operation someone will always have to cover Thanksgiving, Christmas, nights, and weekends. Family and social disruptions remain a difficult issue; society is oriented to using days for work and nights for sleeping. The best approach is for a company to design schedules specific to its plant. Ideally each schedule should be a compromise of employee preferences, operational requirements, and circadian principles. To accomplish this, more attention should be focused on the needs of shiftworkers and the development of policies geared to their special needs, rather than simply adapting day worker principles to shiftwork. For example, what is the value of ten paid holidays per year when you're scheduled to work seven of them?

Among those whose round-the-clock alertness is indispensable to society are police officers. Here is Officer Kerry Day's description of his experience as a shiftworker, presented in his testimony before Congress.

> In 1974 I transferred to the PG County Police and encountered a shift plan in which I again rotated between the three standard shifts, except that the shifts now rotated weekly. I immediately found this to be more tiring and more difficult than Baltimore's monthly rotation system because it required constant changes in eating and sleeping habits without ever adjusting to any of them.
>
> As time passed I became more affected by these weekly rotations and I now consider them to be the single most distasteful aspect of my career, greater even than the frustration of seeing justice sometimes denied, of witnessing various cases of human suffering, and the very real personal danger my job involves. In the eight years I worked weekly rotations I noticed a substantial increase in my levels of anxiety and fatigue. Specifically, I have encountered the following problems:
>
> 1) Strained spousal relationship. My wife initially was very resentful of my constantly changing sleeping patterns and often complained bitterly about a lack of routine and felt that I should sleep with her at night and spend my days

off with her. Eventually she adjusted, but we both agree that the fewer hours we've had to share and the arguments over shiftwork have robbed us of a closeness we might otherwise have had.

2) Fatigue and anxiety. During the past three to four years I have noticed an increase in these problems which, while I'm unable to say definitely resulted from rotating shifts, is nevertheless something which I feel has at the least been aggravated by the shiftwork.

3) A vague sense of time disorientation. This might involve a friend asking whether I'd like to go to a ball game on Wednesday night three weeks hence. I would be unable to give an immediate answer, needing first to make mathematical calculations or pull out a calendar to determine whether I'm working, or sleeping, or free for leisure activities. This lack of stability can be disquieting and frustrating at times.

4) A sense of being different. I am acutely aware that I am in a special category regarding my relationships with others. I am aware that family and friends frequently make special considerations and go to great lengths sometimes in order to include me in certain social functions, trips, etc. Further, I realize that I am left out of other events because of my unusual work hours. I even feel a quasi-guilt over being such a problem. Also I am aware that some people do not really understand the problems peculiar to my shifts and that they regard me as "strange" because of my unusual habits and lifestyle.

5) Eating and sleeping patterns are constantly changing. When working the three-shift weekly rotation, meals became a catch-as-you-can situation, and there was no such thing as "breakfast" or "lunch." A meal was simply whatever was available.

The "normal" world also placed demands upon me which interfered with my ability to adjust sleeping patterns to fit my shifts. Such things as off-duty court appearances, babysitting, doctor's appointments, etc., sometimes occurred during hours I would normally be sleeping. These conflicts often result in situations where I am forced to go 24 to 36 hours without sleeping, and the fatigue and loss of alertness and reflexes has been a source of concern to me, considering the life-and-death nature of police work.

6) Difficulty in maintaining an exercise program. It is very difficult to maintain an exercise routine for any sustained period of time. This may be a serious problem, given the sporadic need for explosive power and stamina which policemen experience in performing their duties, and considering the sedentary nature of the work and its high levels of stress.

There are other disadvantages, such as my children's confusion over whether Daddy's coming or going.

In mid-November 1982 I was placed back into a two-shift plan (I now work a week of day work and a week of evenings, etc.). I would normally consider that a superior shift plan to the three-shift rotation, but my body has not yet grasped the fact that I've changed schedules, and it resents abandoning the system it has known for the past eight years. Despite this, I have noticed an easing of anxiety and I have a more positive outlook since the change occurred.[9]

### Health and Shiftwork

Because complaints of poor health are common among shift-workers, we can assume that they use benefits at a greater rate than day workers. Yet I have found that inevitably, at some point in the course of an educational training session with shiftworkers, a fifty- to sixty-year-old worker, often with twenty to thirty years' job tenure, will slowly stand up and politely say, "What you've said is interesting but I've never had any problem sleeping or adjusting to this backward rotation shift schedule." Some of these old-timers may be among the few people whose inherent biological clock enables them to make an easier adjustment to any type of schedule changes. Probably because the more typical individual has dropped out of shiftwork, the old-timer represents a survivor group. Conversely, the old-timers may be so chronically sleep-deprived that they no longer realize that they are running below par.

The significant differences among individuals in their tolerance for shiftwork may be due to properties of their biological clocks. Several European researchers, for example, have been able to identify two body-temperature patterns that distinguish good and poor shiftwork adapters. Those with a relatively small swing between their temperature maximum and minimum (morning types or larks) seem to have more trouble adjusting to shiftwork than those with a relatively large swing (evening types or owls). To select those most likely to cope well with shiftwork, European researchers advocate physiological studies of job applicants and evaluation of chronobiological questionnaires to be filled out by these applicants. In practice, it may be discriminatory by American standards to apply such physiological tests to the selection of shiftworkers or any other type of workers.

The best-documented health consequences of shiftwork are sleep disorders and disorders of digestion. Averaging the results of nearly 200 surveys conducted over 25 years, we find that 62 percent of 4500 shiftworkers studied complained of sleep disturbance, as contrasted with approximately 20 percent of day workers. Sleep complaints include difficulties in falling asleep or staying asleep, waking up too early, difficulty in waking up, and poor quality sleep. The shiftworker will normally attribute these early awakenings to environmental stimuli—phones ringing, children crying, traffic noises. Although these stimuli may play a significant role, poor work schedules themselves are the primary cause of insomnia. Most shiftworkers average about 5 to 6 hours of sleep while working night duty. Their sleep episodes are shorter because the body's internal alarm clock is trying to wake the day sleeper up. However, shiftworkers who sleep in the daytime for two to three consecutive weeks can obtain a normal amount of sleep (8 hours) despite these noises. This is because the biological cycle adjusts: Daytime is set for sleeping and the alert signal is set for night.

During World War I, an unusually high incidence of stomach disorders occurred among the men and women forced into shiftwork to keep the armaments industry running continuously. The risk of stomach ulcers was eight times greater for shiftworkers than for those on day schedules. Recent studies do not reveal these dramatic differences, probably because those who cannot cope well with shiftwork are usually free to switch to day jobs. About 15 percent of newly hired shiftworkers do, however, develop gastrointestinal complaints, most commonly gastric and peptic ulcers and gastritis.

Shiftworkers commonly believe that they have a shorter life span. Although circadian rhythms are certainly disrupted by shiftwork, there is no clear evidence implicating any long-term health effects. In a carefully conducted study comparing approximately 4000 shiftworkers with 4000 day workers and 500 ex-shiftworkers, no significant differences in mortality rates were noticed over a ten-year follow-up period. These results must be regarded with caution, however; had the 500 remained in the shiftwork category, and another 10 percent not dropped out of shiftwork for "medical reasons," the figures might have been different.

Animals, unlike humans, do not have a choice of shifts. Dr. Franz Halberg, who coined the term "circadian," studied the life spans of 200 mice. One group lived on a normal schedule of lights-on for 12 hours and lights-off for 12 hours. For another group, the scheduled light and dark periods were reversed every week. This is equivalent to rotating from a 12-hour day shift to a 12-hour night shift—a fairly common schedule in the United States—or to crossing 12 time zones every week. Of course, the study does not exactly simulate shiftwork, since the animals' activity is spontaneous rather than controlled. There were no other experimental manipulations. The study was completed when all the mice died of natural causes. The survival time in the mice on the regular schedule was 6 percent greater than those on the shifted schedule. Translated into the human life span, this would be about 5 years.

A recent study of Swedish policemen points up the relationship between longevity and abnormal schedules. As in most police departments, officers in Sweden must cover 24 hours a day. They habitually rotate counterclockwise every week to earlier hours—against the natural drift. A team of Swedish researchers reorganized the shift schedules of 45 policemen to work 3 weeks on the old schedule and 3 weeks on a new clockwise, or phase-delay, schedule. Baseline levels for coronary risk factors (systolic blood pressure, lipids, glucose, uric acid, catecholamine excretion rates) were assessed before and after the trial. The results showed a significantly lower risk for coronary heart disease on the new phase-delay direction of rotation.

The direction of rotation may well affect even straight-day workers, the majority of our society. Recently two German scientists set out to determine whether the switch to daylight saving time increases the number of personal injuries and accidents. In May of 1980, the Federal Republic of Germany for the first time moved its clocks one hour ahead of Central European time. Moving the clock forward is equivalent to a one-hour counterclockwise phase advance. For example, a normal bedtime of 11:00 P.M. will occur when the clock reads 10:00 P.M. on daylight saving time. A phase advance of one hour causes a 23-hour day, which is opposite to the direction of our free-running biological clock. In the week following the introduction of Central European summer time, accidents increased significantly beyond those recorded for the same week in the previous year.

Other studies have compared the rate of traffic accidents during the spring change to daylight saving time (counterclockwise, phase-advance) versus the autumn return to standard time (clockwise, or phase-delay). Traffic accidents during daylight saving time are a good measure of the effects of circadian adjustment. An 11 percent rise in traffic accidents was found in the week following the change to daylight saving time. During autumn changes to standard time, performance studies actually showed some improvement.

Shiftwork is here to stay. For the 25 percent of shiftworkers who prefer their schedules of rotation through day, evening, and night shifts to a normal day schedule, that's fine. Shiftwork does indeed have certain advantages: Being off in the middle of the week means that you can use the ski slopes, go to parks, take in movies at hours when the rest of the town isn't waiting in line to buy tickets. You can participate in your children's activities and visit their schools. Shiftwork also offers premium pay and overtime opportunities for the employee who needs more money fast. An extra bonus is the special camaraderie that shiftworkers develop with each other.

Employees who are able to accept shiftwork as a way of life instead of fighting it are most likely to seize upon its advantages. Still, the shiftworker is confronted by conflicting signals. The work schedule demands alertness at 3:00 A.M., but his family may demand that he be wide awake at 8:00 A.M. and also at 3:00 P.M. Rotating across time zones, blue-collar jet lag takes its toll, just as traveling across time zones takes its toll. But jet lag poses another set of problems and solutions. The same time cues that disrupt travelers—meal times, light-dark changes, and scheduled social events at the new destination—can also provide them with helpful clues for resetting their inner clocks and synchronizing to their new environment.

# 4

# Jet Lag

*If God had intended man to fly east and west, he would have made the equator run north and south.*

—Anonymous

The final loading call for a New York to Rome flight comes over the loudspeaker. A middle-aged couple boards the plane at 8:00 P.M., in time for a late on-flight dinner. About three hours after takeoff, at 11:00 P.M. New York time, the attendants turn down the lights and the passengers try to nod off, but the cramped seats and noise make for uneven sleep. Four hours later, at 3:00 A.M. New York time, passengers are awakened to fill out passport control forms. Lights are turned on and window shades are raised, revealing a bright sun shining over continental Europe. Breakfast is served but little is eaten.

By the time the American couple have gone through customs and checked in at their hotel, it's 1:00 P.M. in Rome, but 7:00 A.M. New York time. Although they disembark in a mood of eager anticipation, they start to feel sleepy after arriving at their hotel. They are eager to see Rome, however, and go on a brief tour of the city. When they return to the hotel, fatigue overtakes them, and they decide to take a brief nap before dinner. Their sleep is interrupted by the maid's knocking to announce that she has come for the morning cleaning. To their surprise, it's 8:00 A.M.—they have slept for twelve hours! The hotel managers are not surprised. Many of their guests lose their first day in Rome.

This incident illustrates many of the features of "jet lag"—fatigue, irresistible sleepiness, unusual sleep-wake schedules. This cluster of symptoms occurs primarily because the internal circadian rhythm is out of phase with the normal temporal environment, creating de-synchronization and resulting in sleep deprivation. If your biological system says it is 3:00 A.M., but local time is 12:00 noon, you may find yourself falling asleep at lunch. Conversely, if local time is mid-night and biological time is 10:00 A.M. you will have a hard time falling asleep. Long, confining journeys—say 8 hours in a bus—will not produce these dramatic effects.

Circadian desynchronization was not a problem when sailors first traveled across the Atlantic. The slow rate of travel allowed the body's internal rhythm to adjust gradually and subtly to new time zones. The first transatlantic fliers, however, were aware of the complications.

### Around the World in Eight Days: Wiley Post

On June 23, 1931, Wiley Post, a pilot from Oklahoma, and his Australian navigator, Harold Gatty, attempted to cir-cumnavigate the globe in their 27-foot, single-propeller plane, the Winnie Mae. The European dirigible, Graf Zep-pelin, had recently landed in Lakehurst, New Jersey, 21 days after taking off from the same location, having completed a round-the-world trip. The airplane had not yet established itself as a mode of international travel, and its future as a commercial vehicle was se-riously threatened by the success of the dirigible. Alarmed lest the dirigible, having proven its ability to safely circumnavigate the world, might triumph over the airplane, Post and Gatty undertook their pioneering journey.

Gatty's preparation included the study of worldwide meteorology, aerial navigation techniques, and analysis of previously unsuccessful flight routes. Post, as the pilot, was preparing for a new human adventure—eight days of nearly continuous flying around the world. There were no guidelines to follow. They had to create their own.

Post's training focused on maintaining two behavior patterns: (1) improving alertness and comfort by keeping a blank mind, and (2) controlling his sleep: "I knew that the variance in time as we progressed would bring on acute fatigue if I were used to regular hours. So for the greater part of the winter before the flight, I never slept during the same hours on any two days in the same week.

Wiley Post and Harold Gatty, with their 27-foot, single-propeller plane, the Winnie Mae.

Breaking one's self of such common habits as regular sleeping hours is far more difficult than flying an airplane!!!"[10]

Wiley Post anticipated the effect of jet lag, or what he called "variance in time." Throughout his eight-day, fifteen-stop trip he kept two clocks—his own wristwatch on New York time to monitor his world record attempt, and a clock on the instrument board to show local time. Despite his preflight strategy, sleepiness did play a critical role during the journey. His second stop after crossing the Atlantic was Hanover, Germany. After being greeted by a crowd of reporters, a marching band, and great public acclaim, Wiley and Gatty climbed back into their plane. The news reporters, cheering crowds, and marching bands applauding their takeoff were surprised to see the plane almost instantly turn around and land once again in Hanover.

Wiley and Gatty were so tired that they had forgotten to check their fuel supply! It was on empty. To the amazement of the airport personnel, the world record challengers returned to Hanover airport for refueling and a second but quieter takeoff.

Upon landing in Berlin at 8:30 P.M., Wiley's fatigue was so great that he fell asleep several times in the middle of responding to a newspaperman's questions. Sleepiness also interfered with concentration. There were times when he was unsure whether he was on the ground or in the air: "Things had become very tiresome after a while and my mind had set in a single track, so that when I landed at Chester, for several minutes I was mentally still flying the plane through fog."[11] This illusion persisted while he was a passenger in an automobile!

Wiley and Gatty were ahead of their time. Problems associated with jet lag did not receive wide attention until airplanes became faster and safer, making air travel a more routine method of transportation. Passengers on overseas flights increased from 2.4 million in 1952 to 6.6 million in 1962, and to several hundred million by 1986.

About twenty-five years ago—in the early 1960s—to reduce the flying time of such West-East routes as Copenhagen to Tokyo, international airliners began flying "over the top"— across the Arctic Circle. With this change, an increasing number of passengers developed symptoms of fatigue, leading to the first widespread report of jet lag. Until then, only a few special cases had been reported. A notable one was that of Secretary of State John Foster Dulles, who in 1953 flew to Egypt to negotiate the Aswan Dam treaty. Unfortunately, his meeting with the Egyptian representatives took place shortly after his arrival, and at least partially due to the fatigue of Dulles and his associates, the project was lost to the Soviet Union, initiating a decade of strong Soviet influence in Egypt. Upon his return, Dulles cautioned his diplomatic colleagues not to conduct important meetings too soon after international air travel. Many corporations have echoed this advice, recommending that company executives not schedule meetings until one or more days after arrival.

Virtually all travelers going "over the top" experience a host of strange symptoms: fatigue, disorientation, decreased concentration and performance, disruption of appetite, insomnia, excessive sleepiness, muscle aches. Surprisingly, however, these symptoms are absent in passengers flying long flights from North to South America.

FIGURE 4-1

## Global Travel

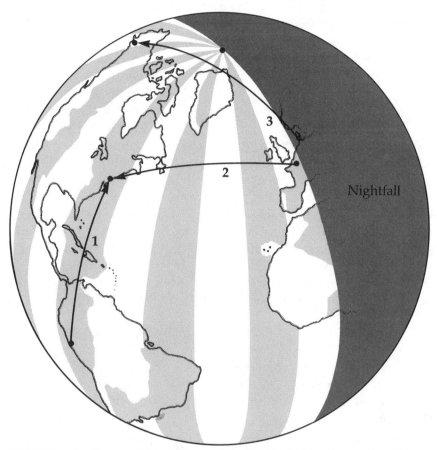

*Flight 1:* Travel in the same time zone does not produce any jet lag, even on a long 12-hour flight from Lima, Peru, to New York. *Flight 2:* Crossing 5 time zones on an 8-hour flight from Paris to New York causes moderate jet lag. Most people will take two to three days to adjust. *Flight 3:* Crossing 11 time zones on a 10-hour flight from Copenhagen to Alaska would cause extreme jet lag. Most people would require a minimum of a week to adjust.

Although a certain amount of fatigue from sitting on a long flight is apparent, this set of intense multiple symptoms does not appear. Also, if you flew completely around the world in 8 hours and landed back at your point of departure you would experience no symptoms of jet lag at all.

The obvious conclusion is that flying extended hours in jet planes does not in itself cause jet lag. The number, rate, and direction of time zone changes are the most critical factors in determining the extent and degree of jet lag symptoms. A 2-hour flight from New York (Eastern time zone) to Chicago (Central time zone)—a single East-West time zone change of 1 hour—is, as one might expect, less disruptive than a 7-hour flight from New York to London—five West-to-East time zone changes. On the other hand, a 12-hour flight from New York to Lima, Peru, with no time zone changes, will not result in any circadian disruption. Fatigue symptoms accompanying such a flight would be similar to the stress of a 12-hour car ride (see Figure 4-1).

The full-blown symptoms of jet lag occur when our internal biological day is out of phase with the new local time. If you fly from New York to London, for example, the clock on the wall in your London hotel may read 8:00 A.M., but your body clock is still on New York time, which is 3:00 A.M. If 8:00 A.M. London time usually finds you fast asleep back in New York, your body will still feel as though it is 3:00 A.M.—hence the ensuing lethargy and decreased performance. Remember, your internal clock tells time based on its most recent experience and will try to keep you on your usual schedule. Even an experienced tour leader has not, by her own account, licked the jet lag syndrome.

> How I envy people who say they don't feel jet lag! I feel it coming and going. For the first week I'm in Europe, I'm exhausted at 3:30 every afternoon, and wide-eyed at 4:00 in the morning. Ghastly. When I get home, I fall into bed at 7:00 P.M. and am wide awake at 5:00 A.M. Lasts about a week. Silver lining: It's incredible how productive you can be between 5:00 and 7:30 in the morning, catching up with all those unpaid bills and unopened mail.
>
> How can I tell it's jet lag? I can't do anything remotely connected with math. Well, it's not my strong suit when I'm in top form, but when I have jet lag, it's hopeless. I've tipped waiters $2.50 (for serving 20 people) and miscalculated the exchange rate by a factor of ten!
>
> Spent the better part of an hour early one morning in Madrid, waiting for hotel rooms to be readied, trying to figure out why there had been a 9-hour time change instead of the 7 hours we expected. It's a complicated problem, even when you're alert. Spain, geographically in a time zone

with England, keeps time with France and Italy (an hour later than England), and besides, they'd gone on "summer" time in early March. I'm not sure we ever did figure it out as we sat in the hotel lobby, trying to be bright and coherent, but managing, at best, to act somewhat stunned.

*—from the diary of a travel agent*

## Variables Influencing Jet Lag

An important variable in determining the degree of jet lag is the direction of travel, not whether the flight is outgoing or homecoming. When traveling in a westbound direction, New York to Los Angeles, for example, we must set our wristwatches and biological clocks back by 3 hours because our day has been extended. (If you normally keep to a bedtime of 11:00 P.M. to 7:00 A.M. and upon arrival in Los Angeles you also stay up till 11:00 P.M. local time, you will experience a 27-hour day. Traveling eastbound, Los Angeles to New York, requires setting your watch ahead by 3 hours, or shortening the day to 21 hours.)

Because our internal biological clock naturally gravitates to a 25-hour day, it makes sense that we can more easily adjust to westbound travel, which extends the day. In a series of studies of jet lag, volunteers have been flown back and forth between Europe and the United States (six time zones) to measure cognitive-motor performance, body temperature, and fatigue ratings. When performance before and after the six-time-zone flight was assessed, it was found that the travelers reached their peak performance within two to four days following westbound flight, but required nine days following eastbound travel. In fact, for two-thirds of the eastbound travelers, the biological clock accomplished the 6-hour adjustment to earlier hours by delaying physiological rhythms by 18 hours. A clockwise change of 18 hours is the same as a 6-hour counterclockwise change. When the biological clock cannot adjust to shorter days, it often will free-run, drifting to later hours until it catches up.

### Rate of Transmeridian Shifts

In 1883 there were 54 time zones in the United States. Baltimore was 10 minutes behind New York but 6 minutes behind Hagerstown, Maryland, and located to its west. Connecticut alone had five time zones. Railroad travelers had to develop expertise in reading schedules and local depot clocks to help unravel their time warp.

In 1884, by international accord, the globe was divided into 24 one-hour time zones and the prime meridian set at Greenwich, England. Because the earth takes 24 hours to make a complete rotation on its axis, 24 time zones was a logical choice. A complicating factor is that time zones narrow as they stretch from the equator to the poles, varying from a width of 1000 miles at the equator to 180 miles at the Arctic Circle. Thus a 12-hour East-West flight from Copenhagen to Tokyo over the Arctic Circle will result in more time zone changes than a 12-hour flight over two cities near the equator. In general, the greatest difficulty in adjustment results from crossing 12 time zones, the least from crossing one time zone. Paradoxically, if you cross close to 24 time zones you'll end up where you started without any jet lag!

A number of other variables may affect the degree of jet lag. Most important of these secondary factors is the influence of *zeitgebers*, or time-giving signals. After a transmeridian flight that crossed six time zones, one group of volunteers was kept inside a test facility for seven days, with no opportunity for exposure to sunlight or to local time cues (indoor activity group). The second group was allowed to leave the laboratory and pursue outdoor activities. This group synchronized to the new time zone in half the time required by the indoor group. Being subjected to local time cues and signals such as meal, work, and sleep schedules, as well as the direct input of light-dark cycles, fosters more rapid adjustment. In this respect, jet lag is different from shiftwork. The night-shift worker has to adjust both to his night schedule and to the diurnal schedule of his family. The jet-lag traveler has the advantage of adjusting to only one new schedule. The lack of conflicting time cues in the new location makes for an easier adjustment.

Other factors underlying jet lag include (1) age—older people have more difficulty; (2) whether one is a morning or evening type person—owls typically do better than larks; and (3) the amount of sleep deprivation associated with preparation for the flight as well as during the flight.

## Air Crews

Just as shiftwork was ushered in by a historical event, so was round-the-clock air travel. Prior to World War II, flying was pretty much limited to daytime hours. The scheduling stresses on air crews were minimal compared to those of today, when crews routinely circumnavigate the earth and fly at night. More recently, with the advent of automated navigation, piloting an aircraft can at times become monotonous and routine.

Leaving New York for Europe at 8:00 P.M. is probably the worst single time from the point of view of air crew alertness. Descent and landing occur at around 3:00 A.M. biological time. Can pilots stay wide awake at 3:00 A.M., the low point on the alertness cycle?

Because you don't see your pilot up in the cockpit—or when you do see him he's wearing an impressive uniform—most passengers develop a feeling of confidence in this legendary figure. In fact, they may believe that pilots are superhuman. Even though passengers cannot help but nod off back in the cabin, they assume that the air crew up ahead is wide awake—as if pilots possess a different physiological system. Pilots fly only about 1000 hours per year, so that it is not the length of the duty period that poses a threat to safety; it is the work-travel schedule and sleep deprivation. Most surveys indicate that the incidence of insomnia and fatigue is much higher for flight crews than for the normal population. Eighty-seven percent reported disturbed sleep. In addition to working schedules that demand alertness at the low point of the biological day and call for sleep at the high point of the cycle, pilots may sleep in a different room every night or day, so that they are constantly subjected to changes in environmental factors—noise, cold, altitude, heat, humidity.

A few years ago a Boeing 707 was scheduled to fly into Los Angeles International Airport shortly after midnight. The flight and crew had originated in New York, so that for the crew it was now about 3:00 A.M., the low point on the circadian alertness cycle. As the plane approached LAX, the air traffic controllers were amazed to see it maintaining its 32,000 foot altitude. Repeated descent clearances issued by the tower were ignored by the three-man flight crew. The situation became more perplexing and dangerous as the 707 overshot the airport and flew fifty miles out over the Pacific Ocean on a dwindling fuel supply. Not until it was 100 miles over the Pacific were the controllers finally able to awaken the three sleeping pilots, who were cruising on automatic pilot, by triggering a series of chimes in the cockpit. At that point the aircraft had just enough fuel to return safely to Los Angeles.

The three men were not poor pilots. This incident is reminiscent of Wiley Post's error in his 1931 flight: Their work schedule did not take into consideration biological clocks. Other near accidents due to sleepiness and poor scheduling have been reported. A jetliner carrying 150 passengers was arriving in Los Angeles early one morning after a night flight from Honolulu. Two hundred feet before touchdown the copilot leaned over to adjust some controls and glanced at the pilot, who was flying the plane. He was asleep. A helpful nudge awakened the pilot, and the plane landed safely. Pilots with whom I have spoken will confidentially admit that fatigue can be a significant problem. Many can relate a specific incident. In an Airline Pilots Association survey of nearly 12,000 members, 93 percent reported that fatigue "was a problem in their type of flying."

The key issue in minimizing fatigue for air crews is no different from that in shiftwork—exact scheduling. At present the scheduling priorities are aircraft availability, passenger convenience (actually local time, not biological time, is considered), and—in last place—crew fatigue. Current Federal Aviation Administration regulations, which have changed little since 1934, permit 8 hours of domestic flying per 24 hours, and 12 hours of international flying per 24 hours. No consideration is given to jet lag or circadian rhythm.

The problems of air crew fatigue are complex and have opened up a Pandora's box for a huge and complex industry. It is difficult to document fatigue as a specific cause of an airplane accident *after* a fatal crash. Fatigue was considered to be a factor in the tragedy at the Canary Islands on March 27, 1977. Two 747s collided on the runway, killing 581 persons. The captain of a

TABLE 4-1

**Crew Schedule Prior to the Bali Crash**

| Day | Destination | S.F. Time | Hours in Flight |
|---|---|---|---|
| 1 | Depart San Francisco | 7:44 P.M. | 5:48 |
| 2 | Arrive Honolulu | 1:32 A.M. | |
| 3 | Depart Honolulu | 3:39 A.M. | 10:56 |
| 3 | Arrive Sydney | 2:35 P.M. | |
| 4 | Depart Sydney | 6:21 P.M. | 7:09 |
| 5 | Arrive Jakarta | 1:30 A.M. | |
| 5 | Depart Jakarta | 2:18 A.M. | 4:22 |
| 5 | Arrive Hong Kong | 6:40 A.M. | |
| 6 | Depart Hong Kong | 4:00 A.M. | 4:30 |
| 6 | Crash at Bali | 8:30 A.M. | |

KLM jet, trying to take off without tower clearance at the end of a long day of flying, reared his plane into a Pan Am jet.

In 1978 a Pan Am 707 crashed in Bali after an experienced crew misread a simple radio direction beam. Ninety-six passengers and all eleven crew members were killed. Their simple performance error may have been due to fatigue, as evidenced by their schedule—not a demanding schedule in terms of air time but extremely disruptive of biological rhythms of sleep and alertness (see Table 4-1).

Air crews on long-distance routes are confronted not only with transmeridian travel, but with irregular work schedules that result in one of the most difficult modes of shiftwork. After several days of flying international routes, the crew may no longer know which is a night flight and which is a day flight. While on duty, they may be consistently out of phase with local time. Disruption of the circadian system of air crews is complex and individualized, depending on the home base and exact work schedule of each crew member.

Although concern about the human factor in airline crew scheduling is growing, most regulations do not emphasize circadian rhythm. A recent survey of regulations in nine countries found that only three took into account disruptions from time-zone transitions or irregular duty hours. Despite the generally disorienting effect of jet lag, there may be occasions when the crew experiencing jet lag will actually exhibit enhanced performance. For example, an American crew arriving in Europe may initially perform better at night than their European counterparts. Unfortunately, these exceptions may foster the mistaken impression that pilots have superhuman powers enabling them to overcome obstacles that we mere mortals succumb to. Pilots themselves attest otherwise. In the words of a U.S. airline captain, "Since it is impossible to store up sleep, it is difficult to prepare your body for these occasional all-night trips. Crew rest regulations are of no help because the crew rest comes after the fact."

Like shiftworkers, frequent transmeridian travelers such as air crews or flight attendants report an increased prevalence of sleep disturbances and gastrointestinal disorders, as well as nervousness. There is also currently no evidence that frequent transmeridian travel and the ensuing circadian adjustment result in a shortened life span. Such studies are difficult to perform and may not be encouraged by pilots on commercial airlines. Pilots receive excellent salaries and time off, which are perceived as partial payment for their potentially decreased longevity.

Concern with aviation safety led the Federal Aviation Administration (FAA) and NASA in 1976 to devise a voluntary confidential reporting system. Pilots, air traffic controllers, and others in the U.S. aviation system are urged to report situations they believe constitute a risk. To encourage responses, a limited waiver of disciplinary action is granted to the pilot in cases where an FAA regulation is inadvertently violated. The crew member submits information about the aircraft, flight times, weather conditions, and other factors that affect safety during the event. The reporter is also free to offer his or her own analysis of any potential danger or actual incident that may have occurred.

Of 20,000 reports submitted in the first four years, 2006 (10 percent) reported incidents involving errors by flight crews—errors in many cases caused by fatigue. These events were most likely to occur during the descent, approach, and landing phases of the flight—that is, toward the end of the duty period. The number of errors related to fatigue increased between midnight and 6:00 A.M.

Several countries have taken circadian rhythm into account in formulating work rules for flight crews. In Germany, for example, air crews flying transmeridian routes are scheduled to return to their home base as quickly as possible. To prevent sleep deprivation when return to the home base is not possible, the airline is required to provide a minimum of 14 hours of rest time and quarters that can be shielded from light and noise during daylight hours. These regulations prevent circadian rhythms from synchronizing to the new time zone and at the same time prevent the fatigue that builds up from chronic sleep deprivation.

Other strategies include a point system whereby each flight is rated for the degree of disruption it causes in the sleep-wake cycle. For example, the number of time zones traversed, the number of successive night flights, and the direction of travel are all rated. The composite score is used to determine how much rest time pilots would be entitled to after a flight. This system also provides the data to make a more equitable distribution of disruptive flight schedules among the air crews. The only exception to an equal distribution would be to avoid assigning older pilots to flights with a high point total. Although no one can adjust his or her biological clock to irregular schedules, younger pilots can better "tough out" the sleep deprivation, fatigue, and dips in alertness.

FIGURE 4-2

## RAF Routes in the Falklands War

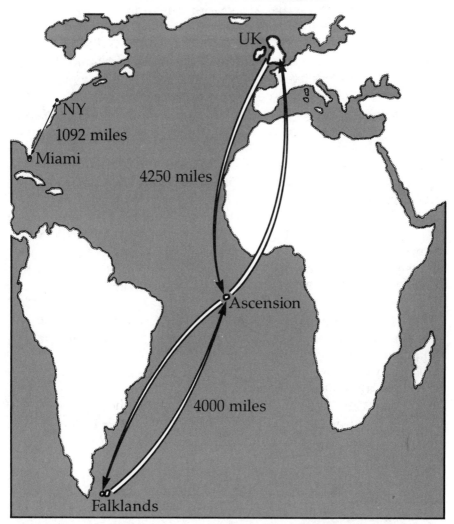

Following the recommendations of a chronobiologist, Royal Air Force crews enhanced their sleep and alertness and were able to carry out military operations for up to 30 continuous hours.

## Conquering Distance: The Falklands War

The South Atlantic conflict over the Falkland Islands in 1982 involved 28,000 British troops, who needed to be supported around the clock, over a distance of 8000 miles, from bases five time zones away. The Royal Air Force was responsible for transporting supplies and troops, as well as for carrying out bombing sorties. One of the major problems encountered was the absence of a forward air field, so that Ascension Island, 4000 miles from the Falklands, served as their closest outpost. A round trip between Ascension and the Falklands took 18 hours or more (see Figure 4-2). The need for round-the-clock air coverage necessitated non-working crews sleeping on board. Sleep difficulties and fatigue were key concerns.

Because several weeks of long missions—up to 30 hours of continuous operation—were envisaged, the RAF relied upon a chronobiologist, Group Captain Anthony Nicholson, to enhance the sleep and alertness of its crews. The primary strategy Captain Nicholson relied on was the use of a hypnotic, *temazepam*. The majority of the air crews used this relatively short-acting medication to get to sleep at various times of the day. Pilots were given a dose 8 hours prior to flight duty and instructed to sleep at that time. With the aid of extra men, strategic use of sleeping pills, and air-to-air refueling, crews were able to maintain continuous operation of 28 consecutive hours and to carry out as many as six long missions within two to three weeks.

The results were excellent. Over 150 support and reconnaissance missions were flown, with only three aircraft lost to enemy action. No pilots were killed and only one was taken prisoner.

The Israeli Air Force has also been concerned with the possible role of circadian rhythm in air force accidents. Researchers from the Israeli Air Force Aeromedical Center investigated all aircraft flight accidents between 1968 and 1980. Only accidents that occurred during peacetime daylight hours and were attributable to pilot error were analyzed. They found that the highest accident rate occurred at 7:00 A.M., one hour after the pilots' 6:00 A.M. wake-up call. After that, accidents declined progressively until the late afternoon, when another peak accident time occurred.

### The Athletes' Dilemma

Fatigue related to travel has become an increasingly greater problem for athletes: Kareem Abdul-Jabbar, star center for the Los Angeles Lakers, in describing an atypical lackluster performance against the Boston Celtics in a Boston-based game, reported that "it must have been the jet lag."

Many coaching staffs are aware of jet lag and circadian rhythm and even offer certain home remedies. Rarely, though, are these based upon scientific research. Coaches often adhere to the traditional "early to bed, early to rise" philosophy. One notable exception was the U.S. men's volleyball team, winners of the 1984 Gold Medal. Doug Beal, coach of the team and a Ph.D. in physiology, applied circadian principles in preparing his team for competition. Because competition was scheduled for 9:00 P.M., practice sessions in the weeks prior to the game were scheduled for 9:00 P.M.–11:00 P.M. Athletes were encouraged to stay up fairly late. No curfew was set, and they were encouraged to sleep late in the morning. Meals and travel were also timed at hours identical with those at which they would be scheduled during the upcoming Olympic Games. "The feedback from the players was extremely positive. It was a help to simulate our game conditions, and the players understood the reasons for it," Coach Beal reported. "Olympic scheduling isn't a simple matter," he writes: "Who you play, when you play, and when you don't are critical considerations."[12]

Prior to the 1984 Olympic Summer Games I had the opportunity to meet with ninety Olympic athletes from the weightlifting team, the men's and women's volleyball teams, and the judo and fencing teams. For these young athletes transmeridian travel was a frequent problem. Most reported that it took an average of two days to adjust to the new time zone. Eight cited specific problems, ranging from "out of balance," "less intense," "fatigued," "sluggish," to "tired, can't play well." The men's volleyball team, for example, reported an average of twenty transmeridian trips per year, with the typical trip crossing six time zones. Most of the twelve members of the team felt that jet lag interfered with performance. The fencing team, which frequently traveled to Europe, noted that they had the greatest difficulty with eastbound flights.

Half of all the Olympic athletes I spoke with reported that their coaches gave them advice on coping with jet lag; the other half received no instructions. A wide variety of suggestions have been offered Olympic athletes: "read reprints," "drink fluids," "get into a new time cycle," "rest," "take sleeping pills," "take no naps until bedtime." Many of the suggestions have some potential benefit, but the lack of any specific program is a weakness the United States Olympic Committee is currently addressing. That such a program is needed is attested by the fact that 60 percent of these 1984 U.S. Olympic athletes reported that jet lag and the scheduling of events had a significant effect upon their performance.

The reactions to jet lag or unusual schedules may differ sharply among the players on a team. One outstanding professional hockey player on the Detroit Redwings team is known for being a night owl. In practice, typically scheduled for the morning, his play is often sluggish, and he is criticized for night-clubbing the evening before. However, during games, scheduled mainly between 7:30 and 10:30 P.M., he is at his peak and is one of the team's top performers.

## Space Travel

Even in space, humans cannot escape the influence of the biological clock. Astronauts and cosmonauts have repeatedly complained about poor sleep, sleep loss, and the accumulation of fatigue during their missions. The space traveler is confronted by an array of new environmental stimuli potentially disruptive to the biological clock: long working days or missions, absence of recurring 24-hour time cues, shifted sleep and work schedules in comparison to ground control, as well as noise from round-the-clock radio transmissions—not to mention weightlessness and loss of gravity.

Brain-wave recordings of Skylab crew members revealed that some astronauts slept less in space, but most actually slept for the same length of time they did on earth. An advantage of space sleep is that one experiences less body movement and fewer awakenings—probably a result of the loss of gravity.

Space travel frees astronauts from the basic 24-hour alternation between day (light) and night (dark). In orbital flights the light-dark cycle alternates about every 100 minutes. At the other extreme, astronauts on the moon would experience a cycle of two weeks of constant sunshine followed by two weeks of constant darkness. Despite these unusual environments, the few studies done to date suggest that space travelers can maintain a 24-hour day if they follow a regular 24-hour work-rest schedule imposed from earth. If not, their biological rhythms will tend to drift, and different rhythms may become uncoupled, creating internal desynchronization. In order to maximize crew performance it is important that this not occur, and that all crew members be synchronized to the same work-rest schedule. In fact, space stations will probably need to be manned 24 hours a day, and specific shiftwork schedules will need to be developed for space work. Because these space personnel will be free of the 24-hour day-night cycle and other social cues, a 25-hour day may serve as the best foundation for developing a rest-activity schedule.

## Combating Jet Lag

Wiley Post flew across the Atlantic in 16 hours and 17 minutes, eclipsing Charles Lindbergh's 1927 record of 33.5 hours. In 1977, the Concorde flew New York to Paris in 3 hours and 45 minutes, averaging 1350 miles per hour. Increasingly rapid flight will have a paradoxical effect: For most travelers, the quicker you get there, the greater the jet lag. However, for those traveling across time zones for a specific event, supersonic travel will eliminate jet lag. A dinner meeting scheduled in Paris, for example, will correspond with lunch time for the New York visitor. The New Yorker could then fly home in time for his normal bedtime.

On the other hand, individuals who plan to stay several weeks in the new time zone may find that rapid transportation increases their sense of confusion. The standard 747 presently takes 8 hours and 15 minutes to fly from New York to Paris; this is a sufficient period of time for travelers to sleep and to start resetting their biological clocks. Proposed 15-minute transcontinental space flights may increase jet lag. Remember, there is no jet lag if you travel slowly across time zones— by ship, for example. The faster you travel, the more abrupt the confusion between your biological time and local time.

What is the best strategy for conquering jet lag? There is really no single solution that will satisfy each traveler. A number of factors need to be evaluated: itinerary, reason for travel, airline schedules, habitual sleep-wake schedules, etc. Computer availability will enable individuals to input information regarding their scheduling requirements and to access specific recommendations. (Such an approach is currently being developed by a software house in California.) At present, there are several possible strategies we can employ.

A number of suggestions have been offered for combating jet lag. All strategies must, however, allow for the fact that our internal clock has evolved over millions of years, with specific parameters. This internal clock can easily adjust to the gradual changes in daylight hours that occur in the transition from short winter days to long summer days. Suddenly, however—in just the past three decades— increasingly large numbers of our population are being continually subjected to frequent, rapid changes of time. While it may be many years before we find the perfect solution, there are several techniques that can provide a measure of relief from the discomfort and disorientation of jet lag.

## Behavioral Approaches

### Arriving Early
Anyone traveling across time zones for a specific event will find this technique useful. By arriving several days early, the traveler will undergo and recover from jet-lag symptoms prior to the event. Following Dulles's experience (and advice), several U.S. presidents have routinely arrived in Europe a few days in advance of summit meetings with Soviet leaders. Arriving early does not eliminate jet lag, but acknowledging its inevitability and setting aside time for it leaves you free, in body and spirit, for the main event.

### Staying on Home Time
By staying on home time rather than attempting to adjust to the new time zone (local time), the traveler avoids disruption of the circadian timing system. The difficulty with this approach, of course, is that you impose your time schedule on others. Your hosts may not find it convenient to let you sleep from 2:00 P.M. to 10:00 P.M. and to schedule dinner parties or meetings for midnight—unless, of course, your status allows you considerable control over your new environment. President Lyndon Johnson, for example, is reported to have maintained local Washington time when traveling aboard Air Force One to confer with Vietnamese leaders. While in Vietnam he maintained his sleep-wake, meal, and meeting schedule on his customary Washington time. The Vietnamese leaders had no choice but to adapt to his schedule. While only a few private citizens are privileged to exercise the same degree of control as the president of the United States, any businessperson traveling from coast to coast can use this approach to advantage. A West Coast representative should aim to start a meeting on the East Coast as late as possible. An East Coast representative should try to start a West Coast meeting as early as possible.

### Synchronizing to the New Environment
A third strategy concentrates on minimizing jet lag once it occurs. After arriving in a new time zone, some people immediately try to change their routines to conform to the new environment. To succeed in this plan it is important not only to adopt a regular sleep-wake schedule based upon local time but also to maintain maximum social

contact. Another effective aid is a short-acting hypnotic drug, such as *Halcion*, to be taken on the first night or two at the new destination. Eating meals on local time, staying out of one's hotel room, and exposing one's self to sunlight are all helpful. This strategy acknowledges jet lag but is successful in reducing the time required to adjust. The couple traveling from New York to Italy, for example, could have avoided their nap and elected to go to bed at 11:00 P.M. Rome time. Forcing themselves up at 7:00 A.M. the following morning to start their day might have helped them make a more rapid adjustment.

## Strategies for Resetting the Biological Clock

Ideally, the way to eliminate jet lag is to reset your biological clock, just as you reset your wristwatch upon arrival in the new time zone. Research has demonstrated that some pharmacological agents can reset the biological clock, although only a few have been successfully used with mammals and none with humans. In lower organisms high doses of caffeine, alcohol, certain protein inhibitors, and a group of drugs called *methylxanthine* are effective in accomplishing this. Some substances foster adjustment to later hours (the equivalent of westbound travel), others to earlier hours (the equivalent of eastbound flights). The effects so far uncovered are complex. Some agents cause shorter days, others longer days; some cause both. All depend for their effect on what time it is in the body when they are administered. Caffeine and alcohol are among the compounds that have proved most effective in resetting circadian clocks in lower organisms. These are consistently offered passengers on transmeridian flights, but not in the higher doses used in animal studies, which, while effective, induced many unwanted side effects. Much more controlled research will be needed to develop specific and safe resetting effects, but we may eventually be blessed with a jet-lag pill.

### The Jet-Lag Diet
Some researchers have advanced the suggestion that special diets may reset the biological clock, but as yet there is little controlled research to support this claim. The most popular jet-lag diet has been derived primarily from research on rodents. In rodents, the timing of meals can induce shifts of biological rhythms, but these findings have not been documented in humans. Another key premise of this plan—that a high-protein diet promotes alertness, while a high-carbohydrate meal stimulates sleep—is also not validated in

humans. Save for personal testimonials, there is little validation of any jet-lag diet.

The U.S. Army tested the impact of jet-lag countermeasures in field studies of troops being deployed from the U.S. to West Germany, a 6-hour time zone difference. The test flights originated in the Midwest at midday and arrived in Germany early the next morning, local time. The test group was not only maintained on a special jet-lag diet but was also restrained from napping until bedtime. At 10:00 P.M. U.S. time each serviceman was given a sleeping pill, and at 11:00 P.M. lights in the airplane were turned off for 5 hours and all the troops were instructed to sleep.

The control group, flying on another plane, was allowed to follow a normal airline routine, with no particular sleep schedule, no sleeping pills, and the usual airline food—which might be considered by many a rather dubious advantage. While the experimental group did adjust slightly better than the control group, it was probably the sleeping pill and improved sleep conditions that were effective. None of the physiological measurements showed evidence that a 6-hour adjustment of biological time had taken place.

*Manipulating Your Schedule*

Another way to reset the circadian clock is to anticipate and adjust the sleep-wake rhythm. This approach requires some effort and scheduling but conforms to the body's limitations: For example, a traveler going from New York to Peking is facing an 11-hour time zone transition. One traveler, over whom I had some influence—he happened to be my father—was successful in manipulating his schedule. By living progressively longer days, he was able to travel to Peking without experiencing any jet lag. Gradually delaying his

TABLE 4-2

**NY to Paris Flight: 8:00 P.M. (NY)–8:00 A.M. (Paris)**

| Sleep Hours | NY Time | Paris Time |
|---|---|---|
| Customary | 11:00 P.M.–6:00 A.M. | 5:00 A.M.–NOON |
| Night Before flight | 10:00 P.M.–5:00 A.M. | 4:00 A.M.–11:00 A.M. |
| *On Plane | 9:00 P.M.–3:00 A.M. | 3:00 A.M.–9:00 A.M. |
| *1st Night in Paris | 7:00 P.M.–3:00 A.M. | 1:00 A.M.–9:00 A.M. |
| 2nd Night in Paris | 6:00 P.M.–2:00 A.M. | MIDNIGHT–8:00 A.M. |

*Use Sleeping Pill*

sleep-wake schedule enabled him to synchronize to Peking time. His first day in Peking he was on the streets of the city observing and participating in early morning exercises. The other members of the tour group, many of whom were considerably younger than he, took up to four or five days to adjust.

Eastbound travel is a more difficult challenge. It is not practical to gradually move your internal clock to earlier hours, as your biological clock can only push about one hour earlier per day. One technique that I use is to take advantage of red-eye specials. Flying from New York to Paris, for example, involves a 6-hour transition to earlier hours. My own strategy is outlined in Table 4-2.

To successfully carry out this plan, I would perhaps take a sleeping pill at 9:00 P.M. to help me sleep, since I am probably not biologically set for sleep at that hour. I actually did this on a ski trip to Switzerland, as a result of which I slept through the entire flight, despite the serving of meals at 3:00 A.M. New York time, the noise, lights, and movement. I had no problems skiing but was fairly exhausted for two to three days. I had diminished the effects of sleep deprivation, even though my biological clock took several days to catch up.

*Toughing It out with Drugs*
A carefully planned regimen of stimulant and/or sedative medications is effective in overcoming jet lag but does not directly attempt to adjust the biological clock. Sleep periods are enhanced by sedatives and wake periods by stimulants without changing the fundamental sleep-wake cycle. In a recent study at the Stanford Sleep Center, volunteers were first tested on a sleep schedule of midnight to 8:00 A.M. and then shifted to a sleep schedule of noon to 8:00 P.M., designed to simulate the crossing of twelve time zones. Volunteers were divided into three groups: One received a placebo capsule; one, a short-acting sleeping pill; and one, a long-acting sleeping pill. Subjects on the placebo averaged about 5.5 hours sleep on the new schedule, whereas those on sleeping pills slept about 7.5 hours. However, when the subjects tried to be alert on their new waking time—8:00 P.M. to noon (i.e., through the night and morning)—they clearly had not adjusted. Those on the long-acting sleeping medications were sleepier than the placebo group, who, although their sleep was shorter by 2 hours, had no drug hangover. Those on the short-acting medication did best, but they still responded as they normally did at 3:00 A.M.—they were not very alert!

# Recommendations for Transmeridian and Trans-Earth Travel

*Passengers*

The goal generally is to adjust rapidly to the new time zone. Shift timing of sleep, meals, and activity to local time immediately after arrival at final destination. Take it easy during the first 24 hours at the new location and if possible do not schedule any important meetings for two to three days. Use a short-acting sleeping pill to enhance sleep on the first night at the new location. To the extent that you can, anticipate transition to new time zone by adjusting sleep, meal, and activity schedule prior to and during flights.

*Air Crews*

Goal is to remain alert while on duty. Avoid adjusting to new time zone by staying on home time as much as possible. Keep a wristwatch that shows both local and home time. Avoid use of long-acting sleeping pills; these may interfere with subsequent performance. Use naps to limit the effects of sleep deprivation. Plan to rest on the first day following return to home base.

*Airline Management*

Goal is to run business efficiently while maintaining high standards for safety. Consider circadian rhythms of alertness in designing work schedules for crews. Factors to be taken into account include departure times, number of time zones crossed, direction of travel, length of duty periods, number of night flights, and other relevant variables. Return crews back to home base as soon as possible. Provide adequate sleeping quarters if return is delayed. Limit the most demanding flight schedules—consecutive night flights across many time zones— to younger crews.

*Athletes*

West Coast teams should try to travel to East Coast events as many days in advance as possible. East Coast teams should attempt to schedule practices to coincide with West Coast time. A basketball game in Los Angeles that begins at 8:00 P.M. and ends at 11:00 P.M. is equivalent to a game played from 11:00 P.M. to 2:00 A.M. in Boston. Similarly, a 1:00 P.M. starting time in New York is 10:00 A.M. in Los Angeles—an hour at which many players are probably sleepy. It would make sense to arrange for a key member of the team, who doesn't play every day—such as a baseball pitcher—to arrive in the new location before the other members of the team.

*Space Travelers*
Maintain a close to 24-hour work-sleep schedule. If round-the-clock operations are needed, crew members should be provided with a separate compartment for sleep and activity.

There are, of course, times when circumstances contrive to render all this advice useless. To return to the category of traveler with whom we opened the chapter—the ordinary passenger—I confess that I wrote much of this chapter while on an airplane. In the past 48 hours I flew from:

TABLE 4-3

| **A Tiring Two-Day Schedule** | |
| --- | --- |
| **Cities** | **Time Zones Crossed** |
| San Francisco to Chicago | 2 east |
| Chicago to Indianapolis | 1 east |
| Indianapolis to Chicago | 1 west |
| Chicago to Denver | 1 west |
| Denver to Gillette, Wyoming | 0 |
| Gillette, Wyoming, to Denver | 0 |
| Denver to San Francisco | 1 west |

Although I experienced rather small time-zone changes, the fatigue caused by endless connecting flights, carrying luggage, sleeping in strange places (right now I'm on the Denver to Gillette flight)—all this is making me irritable. My eyes are burning, I have a slight headache and a neckache. Travel always causes fatigue and some of the symptoms associated with jet lag. The word "travel," in fact, comes from the French "travailler," which means "to work."

# 5

# Why Sleep?

*After 168 hours of sleep deprivation one of the subjects (RS) suddenly went berserk during a psychomotor tracing task. He screamed in terror, pulled his electrodes off, and fell to the floor, sobbing and muttering incoherently about a "gorilla" vision he was experiencing.*

—From *Sleep: Physiology and Pathology, A Symposium*[13]

Sleep-deprivation experiments are designed to study the function of sleep. By depriving volunteers of sleep, experimenters have hoped to find some major physiological consequences that would explain the function of sleep. Perhaps memory, coordination, or cardiac functioning would deteriorate. Dramatic as it is— and sleep deprivation may produce anything from bizarre behavior to simple irritability—the reaction described in the vignette above does not enlighten us about the function of sleep. Hallucinations following sleep deprivation are of short duration. After a good night's sleep subjects who have experienced these episodes are completely recovered and able to perform well.

What critical function does sleep serve? The obvious answer is that it helps us perform and feel better the next day. But for many sleep researchers, that answer does not go far enough; surely there must be some critical physiological process taking place each night. After thirty years of intensive, highly technical research following upon thousands of years of philosophical inquiry, this basic question remains unanswered.

Behind the search for sleep's raison d'être is a hidden assumption that perhaps we should not accept: that sleep is a distinct physiological state with a specific role. The question is so broad it is actually equivalent to asking, Why do we awaken? or, What critical function does wakefulness perform? The question, Why do we sleep? may not be a meaningful or good one to ask. There may be no answer. Just as the earth fluctuates between day and night, plants open and close their petals, and single cells divide and consolidate, humans fluctuate between sleep and wakefulness. Perhaps we should accept the fact that we live in a rhythmic environment and that it is part of the natural order of things that we should also be rhythmic.

Let us take a look at some of the major theories on the function of sleep. An accepted definition of sleep would be "a state of inertia and loss of consciousness of a temporary nature from which one can easily be aroused." Wakefulness might well be defined as "a state of activity and consciousness of a temporary nature in which one can easily be put to sleep!" Viewing sleep and wakefulness as reciprocal behaviors provides an excellent perspective for analyzing various theories of sleep.

Interest in sleep and dreams flourished in antiquity. Aristotle proposed a nutritional/digestive model to explain the sleep process, the key elements of which are widely believed today. He assumed that a process of internal evaporation, equivalent to food ingested minus excretion, was responsible for sleep. Food was ingested and then evaporated into internal gases; these evaporated substances being warm, they rose to the brain, leading to sleep. This explained the phenomenon of postprandial sleepiness. (Genius operates in different guises. Groucho Marx imagined the circulatory system following a similar route: "The blood goes down from the head to the feet, gets one look at those feet and rushes right back to the head.") After a while, according to Aristotle, the evaporated substances cooled down, leading to wakefulness.

Aristotle's model actually has influenced most laymen's and physicians' view of sleep—that sleep is a restorative process caused by a *hypotoxin* (a substance that builds up and then is depleted, like sand in an hourglass). Many research teams have searched for an elusive hypotoxin, or factor S(leep), which, taken from the bloodstream of one animal and injected into another animal, would cause sleep. The results of such studies have been for the most part negative and unreplicated. Even Siamese twins, who share the same circulatory system, continue to show independent sleep-wake behavior: One may be asleep while the other is in a state of active wakefulness.

This suggests that sleep is not controlled by a single distinct chemical.

Is sleep simply a chance for the body to rest, a time-off period? If this were so, then resting for eight hours should have the same results as sleeping for eight hours—and there should be no need to lose consciousness each day. Subjects who try to rest in bed, awake throughout the night, feel sleepy and fatigued the following day. In fact, to keep them awake during such all-night experiments is difficult. If the purpose of sleep were merely to obtain rest, one would expect these subjects to function well the following day. This does not happen. While a full night's sleep predictably improves performance, a full night of restful wakefulness is followed by some degree of impaired performance. The body makes a distinction between rest and sleep. Sleep is far more important.

Curiosity about the brain's role in sleep was stirred by the virus encephalitis, or "sleeping sickness," epidemic in the 1920s. Von Economo (1925) described two contrasting syndromes, insomnia and hypersomnia (excessive sleepiness), and correlated the presence of each with the degeneration of a distinct brain area. This led to the concept of two distinct brain centers—one responsible for causing sleep, the other for causing arousal. The invention in 1929 of the EEG (electroencephalograph), a non-invasive method of measuring brain-wave patterns, was a major breakthrough; for the first time, scientists could study sleeping animals and humans without awakening them. It was not, however, until the 1950s and the introduction of modern sleep research at the University of Chicago that the EEG was fully applied to the study of sleep states, inspiring a surge of hope that the function of sleep might finally be revealed.

In the twentieth century, direct electrophysiological recordings of different parts of animal brains provided the basis for research groups in the Soviet Union, Italy, and the United States to develop arousal theories. They postulated that sleep is a physiologically normal state and concluded that emphasis should therefore be on searching for the brain centers that cause wakefulness. Subcortical brain centers were identified which, when electrically stimulated, caused sleeping cats to awaken, thus confirming that the brain plays an important role in regulating sleep and wakefulness. The stimulation of other brain areas appeared to cause cats to go to sleep. (Because cats sleep about 60 percent of the 24-hour day, these studies have to be carried out under careful controls.)

Although these studies confirmed the importance of the brain as the control center for sleep, the function of sleep remains an enigma. If we think of movies like *2001: A Space Odyssey*—which depicts the dawn of civilization in its opening scene of ape-men crowded around a nighttime campfire—perhaps the assumption that sleep evolved to serve a critical function makes more sense. It may have been a strategy to conserve energy and avoid risk as homo sapiens evolved in a dangerous environment. Even today, in our civilized society, many people feel unsafe out on the streets at night and dependent upon the safety of home and hearth. If we think of sleep as a protective device, then we can assume that dreaming served as a periodic arousal mechanism to sense potential danger during sleep.

Or perhaps sleep really allows for the consolidation of memories, for protein synthesis, for the development of neurological pathways, or for a host of other physiological functions. Each of these hypotheses is advanced by sleep researchers. The problem with all such theories is, Why should we need to lose consciousness and sensory awareness to achieve these gains? "If sleep does not serve an absolutely vital function, then it is the biggest mistake the evolutionary plan has made," says Dr. Allan Rechtschaffen. To test the assumption that sleep serves a vital function, Dr. Rechtschaffen performed an interesting study with pairs of rats living on a disc (like that on a record player). One rat was prevented from sleeping; whenever the EEG showed it becoming drowsy, the disc would start to rotate. This forced the rat to keep moving on the rotating disc—the equivalent of continuously walking up a down escalator. The other—the control animal—encountered the same amount of disc rotation, but was able to catch some sleep when the experimental animal was awake. In this protocol, despite ample food and water, the sleep-deprived animal died within a month; the control animal remained healthy. Autopsies of the sleep-deprived rat showed not only brain pathology but a host of other abnormalities. Certainly sleep serves some vital function, at least in these animals. All the evidence suggests that people also need sleep, though it doesn't tell us why.

No single unifying theory to explain the function of sleep has evolved. Such a theory would have to explain transmission of brain chemicals—why stimulation of certain areas of the brain causes arousal or sedation; neurological function (brain activity center); circadian rhythm; and the psychophysiology of dreaming. We know a little about each of these areas of inquiry, but are as yet unable to put all the pieces together.

On a recent trip to the USSR I was able to confirm my impression that Soviet scientists have had no better luck at unraveling the mysteries of sleep. Here is their summary statement presented at a Soviet sleep symposium in 1982: "Sleep is a process of regulation of the metabolism and the temporal interrelationships in activities of structures and functional systems."[14]

## Sleep Deprivation

A logical approach to determining the function of a behavior is to eliminate that behavior and observe the outcome. To determine the function of sleep, therefore, it seemed logical simply to deprive humans of sleep and chart the results. The fact that sleep deprivation had been used as a method of torture during the Spanish Inquisition suggested that the outcome would be disastrous. One of the first documented studies of sleep deprivation occurred in 1896. A team of research psychologists set out to determine what would happen to sleep-deprived humans under careful observation. Three men were kept awake for 90 consecutive hours (just 6 hours short of four days). All three showed decreased concentration and performance on tests and all experienced hallucinations, but after 12 hours of recovery sleep, all symptoms disappeared.

One difficulty experimenters face in sleep-deprivation studies is the tendency of subjects to fall asleep. As the hours pass, the urge to sleep becomes increasingly irresistible. In 1935 one subject was deprived of sleep for 231 hours and kept awake by being told to push a time clock every 2 minutes. This subject reported severe hallucinations. In 1959, Peter Tripp, a disc jockey, stayed awake in a booth in Times Square, New York, for 200 hours. Using the latest advances in sleep research technology, researchers conducted complete physiological monitoring throughout his stint. Tripp experienced severe hallucinations and paranoia after four to five days, but also completely recovered, with no long-term effects, after 13 hours of recovery sleep. Physiological monitoring revealed multiple brief sleep episodes, lasting 2 to 3 seconds, intruding into the background wakefulness pattern. These brief bursts of sleep, or "microsleeps," make it nearly impossible to completely deprive humans of sleep.

The behavior of sleep-deprived subjects is not actually psychotic. Their hallucinations are usually tactile or visual and short-lived. Psychotic individuals typically experience chronic auditory hallucinations. The *Guinness Book of World Records* cites Mrs. Maureen Weston of Peterborough, England, as the world-record holder for sleeplessness. She reportedly went without sleep for 449 hours (18 days, 17 hours) during a rocking chair marathon in 1977. She, too, experienced hallucinations toward the end of her bout but no lasting aftereffects.

None of the carefully monitored human subjects has died or experienced long-term health problems from prolonged sleep deprivation. Common findings after 100 hours of sleep deprivation include not only hallucinations and paranoia (the likelihood of these reactions is dependent upon both the duration of the deprivation and the subject's psychological profile), but also increased appetite and sexual drive, severe sleepiness, decreased performance, and microsleep. Other effects include increased sensitivity to pain, drooping eyelids, difficulty in focusing the eyes, and sometimes slight hand tremors. Motivation is lost. The problem is not so much an inability to perform, but rather that the willingness to attend to even a simple task declines. Above all, the subjects feel fatigued and "lousy." Typically there is complete recovery following one day of make-up sleep. Although there is not a direct compensation for sleep loss, most sleep-deprived subjects sleep about 12 hours and quickly go back to a normal 8-hour schedule. Surprisingly, performance does not necessarily deteriorate under sleep deprivation. Dr. William Dement reported that during an eleven-day sleep-deprivation study the seventeen-year-old subject was always able to beat him—and he was well-rested—at pinball. Although in the course of counting from 1 to 100 the subject may interject an odd comment or temporarily lose the sequence, the ability to perform does not change. What changes is the motivation to perform: As sleepiness increases, it takes greater incentive to get the subject to perform at capacity.

Sleep researcher Wilse Webb, who conducted many sleep-deprivation studies, described the effects as follows:

> We don't find that the capacity for things like math or playing chess suffers. What's lost is willingness; you would prefer to be asleep. You don't make errors of commission, but omission. Take an example from the war games they do in the Israeli army. One of the prime rules of desert warfare

is that you take every opportunity to top off your water supply, whether you're a tank commander or a soldier with a canteen. Whenever you find water, check your supply and fill it up. Now, as the war games go on over several days and the soldiers miss out on their sleep, following that rule is one of the first things to go—not the ability to shoot or command well.

Take another case—from war games in England. Up north it's cold, and a soldier is supposed to change to dry socks if his feet get wet. Otherwise your feet are going to go out on you, get frostbitten. But as the games go on, they forget, they just don't bother. They can still hit a target as well as ever. It's the low-level, boring routines they skip. You see the same thing with nurses working the night shift: They give all the correct drugs, but they skip walking down the hall to check on a patient.

Low-demand, self-motivated tasks fail. There's a lovely old prayer in the Episcopal Church: "Dear Lord, preserve me, keep me from leaving undone those things that I ought to have done." And that's what goes first: You leave undone those things you ought to have done. If you've got enough inner motivation and drive, you can go that last mile. If not, when you're so sleepy, you just give up. It's not your thinking or memory that goes—it's your will to go on.[15]

The results seem disappointing. As a sleep researcher I tend to hope that sleep serves a vital function. If this is so, then sleep deprivation should produce some dramatic change. Basically, we have found that sleep deprivation makes one sleepy. Instead of being disappointed, perhaps we need to accept this as the bottom line. Unlike the rat, humans are immensely adaptive. Loss of sleep may slow us down but it won't kill us. Señor Izquierdo, whose story is reprinted here, is still alive after 40 years of sleep deprivation!

### Sleep and Health

How much sleep do I need? This has always been an embarrassing question for a sleep researcher. Almost everyone is interested in sleep, and this question is frequently asked of me by any stranger who discovers my peculiar line of

## A Man Who Hasn't Slept Since World War II*
### By Robert Powell

*San Antonio de Los Banos, Cuba*

The only outward sign that Tomas Izquierdo has lived without normal sleep for 40 years is the pair of dark glasses protecting his sensitive eyes.

The former textile worker is a mentally alert and young-looking 53, and he and his second wife recently had a son. Yet a large dossier of medical evidence suggests that he lost his ability to sleep at the end of World War II and has remained awake ever since.

"As far as we know, no case like it has been reported in medical literature anywhere in the world," says Dr. Pedro Garcia Fleites, one of Cuba's leading psychiatrists, who has treated Izquierdo for the past 16 years.

In 1970, Garcia Fleites and a team of doctors at Havana Psychiatric Hospital kept Izquierdo under constant observation for nearly two weeks. Even when he rested with his eyes closed, the electroencephalograms continued to register the brain activity of a person fully awake.

"He has no natural sleep. The nearest thing Tomas gets to sleep is a drowsiness produced by the drugs prescribed for him," the psychiatrist said in an interview.

Like the rest of us who cannot survive more than a few days without sleep, Izquierdo suffers from exhaustion and needs periodic rest.

Even in a state of drug-induced narcosis, however, he is unable to escape completely from the consciousness that has haunted him since 1945.

"I dream just as I would say everybody else dreams. The difference is that I know positively that I am awake and that I am active," Izquierdo said at his home in this small town near Havana.

He and Garcia Fleites have different ideas about the origin of his chronic insomnia. According to the psychiatrist and other doctors familiar with the case, Izquierdo's sleep mechanism was probably damaged by an attack of encephalitis—

an inflammation of the inner brain—when he was 13.

Izquierdo thinks his insomnia derives from a psychological trauma he suffered during an operation to remove his tonsils. A throat hemorrhage sent blood spurting out of his mouth and the terrified adolescent thought he was dying.

The horrific sensation of dying subsequently repeated itself in nightmares, and Izquierdo says that he began resisting sleep to avoid them. According to his own account, within a few weeks he found he had stopped sleeping altogether.

Since then, more than 40 doctors have tried hypnosis, electroshock treatment, acupuncture and experimental drugs to restore Izquierdo's ability to sleep.

Izquierdo has even resorted to spiritual mediums and voodoo doctors, but all have apparently failed to allow him to sleep.

Affectionately known in San Antonio de los Banos as "Tomas who doesn't sleep," Izquierdo used to work double shifts at the local textile factory. He was retired in 1968 on medical grounds when symptoms connected with his inability to sleep became evident.

According to Garcia Fleites, Izquierdo's memory began to fail, and he showed a progressive lack of self-confidence.

Izquierdo himself admits that he now finds it difficult to remember dates or retain the contents of a book.

At present, he passes the time doing odd jobs for friends and neighbors and driving couples to weddings in his immaculately kept 1955 Chevrolet Bel Air.

Transcendental meditation has become his most effective form of relaxation, and on most nights he rests for a few hours from 3 or 4 A.M. onward, meditating or drifting into a drug-induced stupor. Sometimes Izquierdo feels that he can still go happily for several days without rest.

"But some days it is just the opposite," he says. "There are days when I am good for nothing. I feel drained, drained, drained, mainly in a mental sense, but physically as well."

"It's a tragedy," Izquierdo says, "a tragedy within my own self."—*Reuters*

*Published in the *San Francisco Chronicle*, April 22, 1986.

work. When I am on an airplane and don't want to be disturbed I give my short answer: "Seven to eight hours." In a prolonged conversation, my answer is, "Since we do not know the reason why we sleep, we do not know how much sleep you need." These answers never feel satisfying. Another source of frustration is that I know that my fellow passenger knows the reason for sleep—to rest and restore the body—whereas I am thinking about the hundreds of research studies focused on this issue. Like Eskimos, who have 100 different words to describe snow, I can give 100 different answers—which can be an advantage or disadvantage, depending upon whom I'm speaking with. Here we are, after 25 years of intensive research studies, without an answer to the most fundamental questions.

There is no magic number. Daily sleep requirements vary from person to person and change with age. Statistics show that 90 percent of Americans sleep between 6 and 9 hours a day, with the average 7 to 8 hours. But telling everyone to sleep 7 to 8 hours is like telling everyone to wear a size 8 to 10 shoe. Individuals' needs differ.

If you really want to find out what your own sleep requirement is, you can try the following test: First pick a conventional sleep schedule of 11:00 P.M. to 7:00 A.M., allowing for 8 hours of sleep; stay on it for about a week. Then try a 1:00 A.M. to 5:00 A.M. schedule, getting your sleep down to 4 hours. If you feel alert, well rested, and perform well on the 1:00 A.M. to 5:00 A.M. schedule, you can conclude that your sleep requirement is 4 hours.

When we do this kind of study in the sleep laboratory we are able to take accurate measurements of alertness and performance. We find that the requirement for sleep is often *greater* than 8 hours. For many of us, natural sleep length may be cut short as the increasing demands of work and of social and family life impinge upon our natural rhythm. Our daily sleep requirement, however, is probably genetically programmed, and there is not too much we can do to change this biological rhythm.

Many people are aware of their own sleep barriers—or minimum sleep requirements. For example, if I can get at least 5 hours of sleep I feel fine the next day, but with only 4 hours I feel sleepy, and 3 hours will find me fighting to stay awake in the afternoon. The typical threshold is 4½ to 5 hours of sleep. Mood and performance may suffer when that threshold is crossed.

More recently, I have been able to give a somewhat more satisfying answer. "You need enough sleep to be alert the next day," and "We can accurately measure with physiological tests whether you are normally alert." (See Chapter 7 for measurement of optimal and abnormal alertness.) Additionally, I say that sleep length may have an important impact on longevity.

In 1959, the American Cancer Society surveyed a representative sample of over one million Americans in the San Francisco Bay Area to determine the health and lifestyle habits most predictive of longevity and illness. Six years later the respondents were recontacted and mortality rates were assessed. Surprisingly, one of the best guides for predicting which respondents would be alive at the six-year follow-up period was how long they slept. In some age groups males with a history of 4 hours of sleep per day were ten times more likely to have died than those who slept between 7.0 and 7.9 hours. Nor was sleeping too much a good idea: Mortality rates were twice as great for those sleeping 10 or more hours. Even the groups who slept 8.0 to 8.9 hours or 6.0 to 6.9 hours had a greater mortality than those who slept between 7.0 and 7.9 hours. Death from coronary heart disease, stroke, and aortic aneurism were especially prevalent among the long sleepers. The complaint of insomnia was not an indication of greater mortality, but the frequent use of sleeping pills was associated with higher death rates.

Confirmation of the importance of sleep length is provided by the National Center for Health Services Research. Adequate—neither too much nor too little—total sleep time was considered one of the six most critical factors affecting mortality and morbidity. The others were not smoking, limited consumption of alcohol, regular exercise, regular meal schedules, and maintaining the proper weight. The average life expectancy for men aged 45 who follow all six of these good health practices was eleven years greater than for those who followed three or fewer. For women the average life expectancy was seven years greater.

### Less Sleep, More Work?

Thomas Edison believed that every industrious American should be able to spend less time sleeping and more time inventing. He favored the notion that strong will and discipline can overcome biology. Although folklore has it that he slept very little at night but took brief refreshing catnaps, his own diaries indicate otherwise. He scheduled himself for 4 to 5 hours of sleep per night but had trouble waking up on time and would drift back to sleep for an extra hour. He also napped frequently. Edison saw the need for sleep as a holdover from caveman days (he might have been right) and hoped the electric light would enable man to free himself of it. But one hundred years after the invention of the light bulb, most of us are still sleeping away one-third of our lives. Edison was a great inventor, but a poor physiologist!

Many people try to reduce their total sleep time in order to allow themselves more hours of productive wake time. This seems like a simple and obvious solution to a predicament most of us face at least some of the time—too much to do and not enough time to do it— and has been studied by sleep researchers. Most studies of partial sleep restriction are done for a few consecutive nights. The following day, subjects typically show a drop in performance only when engaged in boring, routine tasks—for example, adding a column of numbers. No other major errors are noted. When subjects are reduced from 8 to 5½ hours of sleep for longer time periods, almost every measure of performance remains stable; no decrease in competence is detectable. On routine vigilance tests the subjects responded less frequently, indicating that they were less motivated. Although most subjects reported daytime drowsiness in the first week, this complaint subsided by the eighth week. You might conclude that these subjects would therefore want to stick to their new sleep-restricted schedule. Not true. When the study was over, virtually all the volunteers returned to their customary 8 hours of sleep, because they just "felt better" with more sleep.

Even though they could gain 2½ hours of wake time a day without any noticeable detriment, most people won't reduce their sleep time except in emergency situations.

All the evidence suggests that we have a physiological need to sleep about one-third of the day, that we feel bad when we get less, and that too little or too much sleep may be bad for our health. It's hard to fool Mother Nature.

Only a very few normal persons habitually sleep for a very long time (about 10 hours) or for a very short time (2 to 3 hours) and still function well in the daytime. There are, of course, patients with sleep disorders accompanied by extreme sleep lengths and impaired daytime functions. Most of those who claim to sleep only 2 to 3 hours or 12 hours per night are simply wrong. The "short sleepers," when measured polygraphically for 24 hours, are typically found to nap during the daytime and to enjoy more sleep at night than they report.

The "long sleepers" typically confuse excessive time in bed with excessive sleep hours. In general, true short sleepers tend to be energetic, ambitious individuals who show little evidence of psychopathology. True long sleepers, by contrast, tend to be more neurotic and depressed.

## Napping

Napping is a universal phenomenon, understandably most common among infants, children, the elderly, and shift-workers. In certain cultures, the nap is a fixed part of the day for everyone. In a study of a Mexican community, nearly 80 percent of the adults were found to take four or more naps per week. In general, these naps are added to the normal 7 to 8 hours of nocturnal sleep. The traditional siesta in Mexico and in most societies typically occurs between 2:30 and 5:00 P.M., roughly the midway point between the normal bedtime and waketime. Morning naps and early evening naps are much rarer, suggesting that afternoon sleepiness may reflect a biological rhythm. In looking at the 24-hour day, 2:00 P.M. to 5:00 P.M. and 11:00 P.M. to 7:00 A.M. are optimal sleep times, while 8:00 A.M. to noon and 6:00 P.M. to 8:00 P.M. are rarely spent in sleep.

Habitual nappers fall into three groups: replacement nappers—those who nap to make up for lost sleep or in anticipation of lost sleep; appetitive nappers—those who nap for psychological reasons; and siesta nappers—those whose cultural environment provides a built-in period for naps. Replacement naps tend to last about 75 minutes a day, while appetitive naps keep closer to an hour. Siesta naps are longest, extending to about 90 minutes.

Many famous personalities have claimed that a brief nap of 10 minutes is refreshing, resulting in improved performance and vigor. What is the evidence for these claims? Only a few nap studies have been carried out and none that has directly scrutinized the claims of individuals like Thomas Edison. In those that have been done, however, researchers have been able to demonstrate certain conclusions by awakening napping subjects in order to measure various psychomotor performance tasks. Among their conclusions are these: (1) A sudden awakening from a nap is followed by decreased performance—a phenomenon called sleep inertia, which usually disappears within half an hour after awakening. (2) Napping does not lead to a dramatic performance improvement. Half an hour following a nap, performance will either remain on the same level or show slight improvement.

Other performance issues researchers have considered the optimal length of time and the optimal hour for napping. Preliminary studies suggest that there is little difference between 1-hour and 2-hour naps. The value of naps is affected by their timing in relation to circadian cycles: A nap at 8:00 P.M. may help the night shiftworker on his subsequent shift but may interfere with subsequent nocturnal sleep in a day worker.

Regardless of recent findings, I feel that some of my most enjoyable sleep occurs during a late afternoon nap. Around 2:30 P.M. I often start to feel sluggish. On rare days when I nap for 2 hours, I initially wake up groggy. But around 7:00 P.M. I've peaked and can work effectively until 10:00 P.M. On days when I am unable to nap, I either tough it out or drink a cup of coffee and start to feel more alert around 4:00 P.M. But on those days I'm usually feeling more like watching TV after dinner than doing anything else. This makes the siesta idea sound like a good one, but unfortunately the siesta is being phased out in many cultures—in Italy and Mexico, for ex-

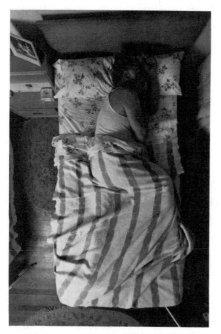

**Is he enjoying a good night's sleep? Yes, this is an example of a typical night's movement for**

ample—because the traffic conditions make it too difficult to commute to and from work twice a day.

## Recording Sleep

We all have electrical activity or currents in our brains. The currents are so tiny that you can't feel anything if you touch your head. However, if you attach electrodes to your scalp, and amplify the low currents onto a polygraph (an ink-writing amplifier) it is possible to obtain an accurate recording of these brain discharges. This is precisely the method used to measure sleep. Fluctuating electrical brain patterns or "brain waves" are amplified, recorded, and deciphered. In addition to the measurement of brain-wave patterns, the monitoring of eye movements and muscle activity allows the sleep researcher to make an accurate assessment of the different states of sleep.

### NREM Sleep: An Idling Brain in a Moveable Body

During *alert* wakefulness the brain-wave patterns have a fast frequency, eyes frequently move rapidly up and down and from side

an average sleeper. (From 2:57 A.M. to 3:48 A.M.)

to side, and tension is evident throughout many muscle groups. During *relaxed* wakefulness—for example resting in bed while awake—the EEG slows down, eyes do not move very much, and muscle tension is absent except in the mimetic (facial) musculature.

The onset of sleep does not turn off the body or force it to become inactive. The body goes into an idling or cruise-control state. Although respiratory rate and heart rate slow down slightly they remain regular. Even the mind is not completely shut off during NREM (Non-Rapid Eye Movement) sleep. If a subject is awakened he will report some mentation about 25 percent of the time. NREM sleep is composed of four stages, which are defined somewhat arbitrarily by the presence or absence of brain-wave patterns. Some researchers suggest there may be many more stages of sleep that need to be described, and others say the concept of stage distinctions is arbitrary and not very meaningful or productive.

The first stage of NREM sleep, appropriately called Stage 1 sleep, occurs right at the transition from wakefulness to sleep. To define this transition, researchers taped open the eyelids of sleepy volunteers. While brain-wave patterns were recorded, subjects were instructed to push a button every time they saw an image flashed upon a movie screen set in front of them. At a certain point, the subjects stopped responding even though their eyes were open and images were present. This moment—when subjects no longer perceived their environment—was the onset of sleep. The brain-wave patterns revealed a characteristic pattern at that point: a slowing down of brain-wave frequency and the presence of certain sharp wave patterns. Furthermore, eye recordings indicated that the eyes roll slowly from side to side. Muscle tension is absent except in the facial and neck muscles.

Stage 1 sleep usually lasts only about 10 minutes. A very light stage—many individuals awakened during this stage will claim that they were not asleep—Stage 1 sleep is accompanied by characteristic mentation, starting out as realistic thinking only to become distorted as we drift asleep: "I was thinking about an exam I need to take tomorrow" (reported just prior to Stage 1 onset), "and suddenly I felt like I was floating alongside a piece of paper. Maybe it was the exam questions" (reported during a Stage 1 awakening). Stage 1 sleep mentation can be vivid, like a dream, but more typically reflects a change from realistic to distorted thinking. Sensations of falling, floating, feelings of anxiety or peacefulness—all are commonly reported upon awakenings from Stage 1.

The amount of Stage 1 sleep is a good indicator of a sleep disorder. In a regular sleeper, it is rapidly completed; poor sleepers, on the other hand, may go in and out of Stage 1 sleep and wakefulness, plagued by repetitive thoughts. Nevertheless, most of us quickly make it to Stage 2 sleep, which is still a fairly light sleep marked by specific brain-wave patterns. Slow-rolling eye movements of Stage 1 come to a halt. In this stage, sleepers will respond to certain noises: A sleep technician who makes a noise in the control room will find that a particular brain-wave pattern shows up on the polygraph: The sleeping brain registers a response; eye movements are absent. Even though 50 percent of the night may be spent in Stage 2, there is probably less research on this stage than on any other.

After about 20 minutes of Stage 2 the sleeper enters a truly deep stage of sleep, which we call Stage 3-4, because it is difficult to separate the two. At this time all the cortical brain cells appear to be firing synchronously, resulting in large, slow waves on the polygraph. Stage 3-4 occurs primarily at the beginning of the sleep period. That is why if you get a phone call about 45 minutes after you've gone to bed, you're likely to be very confused. It can take up to 15 seconds to really wake up from a Stage 3-4 sleep. The effect is even more pronounced in children.

If you try to awaken a child during Stage 3-4 sleep he or she is likely to be dazed and will not remember being awakened the next morning. Arousals out of Stage 3-4 sleep tend to place children in a state midway between sleep and wakefulness. In fact, this is the time of the night that sleepwalking, enuresis (bedwetting), night terrors, and sleep-talking are most likely to occur. Waking a 10- to 12-year-old during Stage 4 sleep is a likely way to trigger an episode of sleepwalking in a normal child. Stage 3-4 sleep may have restorative value. Following the end of sleep deprivation studies, it is the first stage to be made up. This stage, which accounts for 25 percent of total sleep time in children, declines slightly in young adulthood, and then diminishes further in middle age and older years. In elderly subjects, Stage 3-4 sleep is often absent. The decrease in this deep slow-wave sleep is probably the result of a natural aging process in which certain brain cells cease to function.

## REM Sleep: An Active Brain in a Paralyzed Body

Stages 1–4 comprise the four NREM sleep stages. After about 90 minutes of NREM sleep the first REM (Rapid Eye Movement) period begins. The start of a REM period is marked on the polygraph by a drop in facial muscle activity, a change in the EEG to a brain-wave pattern similar to wakefulness, and the presence of rapid, conjugate eye movements. In fact, the polygraphic tracings look so much like wakefulness that some researchers call it paradoxical sleep. In fact, one of the more difficult distinctions to make in interpreting a sleep recording is that between wakefulness and REM sleep.

The discovery and understanding of REM sleep by the Chicago group (Aserinsky, Kleitman, and Dement) in 1953 led to a tremendous interest in sleep research. Prior to this discovery, sleep was considered to be an off-state of the organism, not worthy of study. Besides, scientists did not relish the idea of staying up all night, and, finally, it was difficult to study sleepers without awakening them. Looking back from 1986, the discovery of REM sleep did not fully unravel the mysteries of sleep and dreams, but it did lead to the discovery of a number of important sleep disorders which, with modern treatment, are improving thousands of patients' lives.

REM sleep is a distinct state of sleep or consciousness. Heart rate and respiratory rate are irregular, often greater than during wakefulness. Eyes move rapidly from side to side. Muscle twitches are abundant, but spinal reflexes are absent. The body is immobilized, unable to act out the dreams being experienced. According to reports of research subjects upon awakening, dreams occur 80 percent of the time one is in REM sleep. The discovery of REM sleep and its association with dreaming was a major breakthrough in sleep research. How could sleep be merely a restorative, recuperative process if it included a state of intense organismic excitation? The discovery confirmed that sleep was not merely an off-state—and that it might be worthy of study. The entire field of sleep disorders medicine sprang from this remarkable finding. A variety of physiological parameters routinely studied during wakefulness were now for the first time studied during sleep, and the ability to detect sleep abnormalities accordingly enlarged. During REM sleep oxygen consumption increases and more blood flows to the brain than during wakefulness;

FIGURE 5-I

## Sleep-Stage Cycles

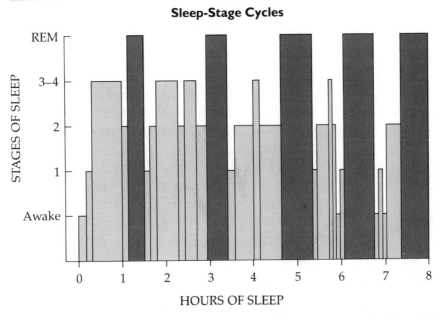

Distribution of hours of sleep for a typical young adult. Stage 3–4 sleep predominates in the first half of the night, REM sleep in the second half. A new REM cycle starts about every 90 minutes.

blood pressure, heart rate, cardiac output, and respiration rates increase and are more variable than during wakefulness.

As the night progresses REM periods become longer. This explains why there is a good chance that when you wake up in the morning you will be in the midst of a REM period and will be able to recall a dream. There are typically four to six REM periods per night, each lasting anywhere from 10 to 45 minutes and constituting about 20 to 25 percent of the total night's sleep (see Figure 5-1).

In 1969 Stanford's was the only sleep clinic in the United States. In 1975 there were five clinics, and in 1986 there are several hundred sleep clinics and hospitals throughout the country that conduct sleep studies by means of recording devices. With technological advances, studying sleep is becoming more complex. In the 1950s, sleep researchers considered the physiology of the subject to be their domain. Sleep was monitored by attaching surface electrodes to the scalp and face to measure brain waves, eye movements, and muscle activity. As the interaction of sleep disorders with other disciplines has expanded, nocturnal recordings have become increasingly complex. Esophageal balloons, oxygen saturation recordings, video fluoroscopy, cardiac catheterization—essentially all the clinical procedures and measurements that doctors normally perform during the day—can now be continued at night to obtain a complete understanding of the body's physiology.

For the patient, most sleep clinic procedures are fairly benign. Even our most anxious patients often fall asleep while technicians attach electrodes to various parts of their bodies just prior to bedtime. (Only a handful of patients are unable to fall asleep during the night study.) A polygraph prints out a second-by-second description of physiological measurement of EEG, eye movements, muscle activity, respiration, and heart rate. The data gathered from a single night is massive—up to one-quarter mile of printout paper. Fortunately, sleep technicians can score an entire record in a couple of hours.

*Rapid Eye Movements: A Night at the Movies?*
Prior to Aserinsky's observation of rapid eye movements at the University of Chicago's sleep laboratory, a few other scientists had hypothesized a connection between dreaming and eye movements. In 1892, G.T. Ladd at Yale surmised that during vivid dreams our eyes focus; and Edmund Jacobson, who developed progressive muscle relaxation, suggested that eye movements should be amenable to electrophysiological detection during dreams.

Since Aserinsky's work numerous studies have been designed to further understanding of the role of rapid eye movements. Do the eye movements represent the dreamer's attempt to follow the dream images—as one does in watching a movie—or are they one of many physiological systems activated during REM sleep? Although there is sometimes a close association between eye-movement patterns and subsequent dream reports—for example, a series of horizontal eye movements associated with a dream report of watching a tennis match—such direct correlations are rare. In fact, evidence exists against the notion that REM is a night at the movies. Cats without a cerebral cortex, the part of the brain responsible for thinking and visual imagery, continue to have rapid eye movement during REM sleep. Probably in humans, too, the rapid eye movements are not simply indicative of the scanning of dream content but reflect one dimension of this intense state of physiological activity.

*Age and Sleep Patterns*

The "smile of the angels," an infant's natural smile, first occurs during REM sleep, long before the child is able to smile in wakefulness. In fact, 50 percent—equivalent to 8 hours per day—of infant sleep is REM sleep. In premature infants, up to 75 percent of sleep is REM sleep. Whether these infants are dreaming or not remains debatable. The high presence of REM sleep has led some researchers to suggest that the main biological function of REM sleep is related to neonatal maturation—assisting in the development of the central nervous system. The percentage of REM sleep gradually declines to fill about 20 to 25 percent of sleep time in early childhood and remains fairly consistent throughout the life span.

## Animal Sleep

Little is known about the sleep of non-mammalian species. Fish do show periods of inactivity—usually when they can find protection in a secure hiding place and when enemies are not present. Most fish often sleep by burying themselves in the sand, leaving their mouths and eyes exposed. Although it is unclear whether the perpetual swimmers—certain species of sharks—actually sleep while swimming (I imagine they do since humans can sleep while driving above ground), other sharks certainly sleep. This is confirmed by a diver who was able to grab one species of shark by the tail before it woke up and—fortunately for him—swam away. An electrophysiological study of fish that recorded eye movements, muscle tone, gill movements (respiration), and heart rate found no evidence of REM sleep. One research team observed rapid eye movements during inactive periods in Bermuda reef fish, which suggests that some fish may have brief episodes of REM sleep.

The distinction between activity-inactivity cycles and sleep-wake cycles is less clear in the non-mammalian species. We cannot really be sure that the fish's inactivity should be called sleep, as it does not show the type of brain-wave pattern we use to define mammalian sleep. The sleep of amphibians and reptiles is subject to the same restraints. (While I was doing human sleep research at the University of Chicago, a series of crocodile sleep studies were being carried out. I stayed close enough to find out that crocodiles do show a distinct sleep brain-wave pattern but no evidence of REM sleep.) In general, amphibians and reptiles do not show REM sleep as seen in mammals, but they do show very active sleep episodes, which some

FIGURE 5-2

## Average Sleep Time of Mammals
## Per 24 Hours

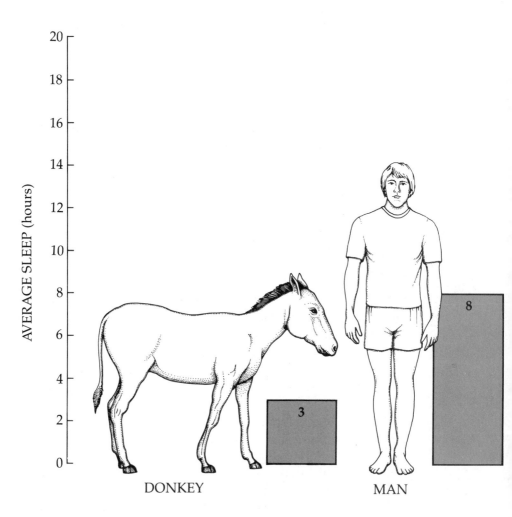

scientists have defined as a rudimentary type of REM sleep. Birds, however, do show brief episodes of REM sleep.

All the mammalian species have REM sleep: a distinct brain wave pattern, loss of muscle tone, and rapid eye movements. So far no clear relationship between the amount of overall sleep per day, or the amount of REM sleep per day, and the mammalian type has been proven. Figure 5-2 shows average amounts of sleep for different mammals.

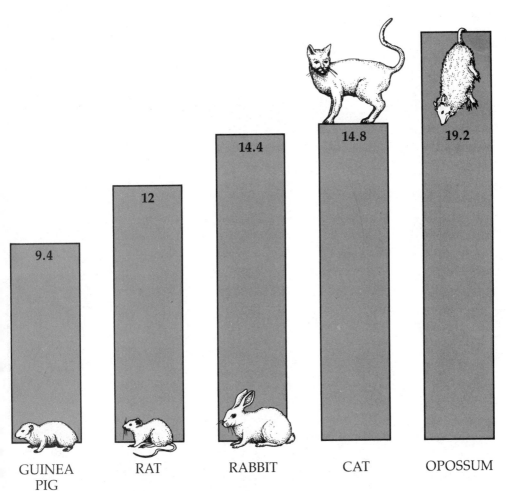

| GUINEA PIG | RAT | RABBIT | CAT | OPOSSUM |
|:---:|:---:|:---:|:---:|:---:|
| 9.4 | 12 | 14.4 | 14.8 | 19.2 |

We may never find out why we sleep, but we do know that we will continue to sleep. And we also know that we spend a portion of our lives in the state of existence called REM sleep—and that we dream while we sleep. Why does our biological system schedule us for a daily period of dreaming? What is going on during these mysterious hours beyond our control? Let's take a look at some of the answers.

# 6

# Dreams

*Sometimes a dream about a banana is a dream about a banana.*

—Overheard at a Meeting of Sleep Researchers

an dreams predict the future? In 1972, while undergoing psychoanalysis, I learned to improve my dream recall in order to bring to my therapy sessions data that would enable me to probe my unconscious. I was living in Chicago at the time and was an avid fan of the Chicago Black Hawks, the area's professional ice hockey team. One night I dreamed about the outcome of the seventh and final—the deciding—game of the Stanley Cup ice hockey championship finals between Montreal and Chicago, which was to be played the following week. I dreamed that Montreal would win the final game 3 to 2 and that Henri Richard, a proud but aging star, who had recently been benched and was nearing the end of his career, would score the winning goal on a backhand shot. In the therapy session my analyst and I discussed the various meanings and associations of the dream, focusing on themes of competition, aggression, identity, and so forth. The reason I remember the dream, however, is that the following day Montreal beat Chicago 3 to 2 to win the Stanley Cup. Henri Richard scored two of the three goals, including the game-winning goal.

Being a sleep researcher and scientist, I tried to rationalize my experience: 3 to 2 was a common score; Richard, although not a big goal scorer that year, was a clutch performer. Perhaps my dream was simply the most logical sequence. Dreams of such seemingly uncanny foresight are uncommon; yet to experience one or even several in a lifetime is not unusual.

Years later, however, a new approach to understanding this predictive dream was suggested by an article I came across in a scientific journal. It was based on probability theory: Approximately 20,000 people were at Chicago Stadium to see the game, and I estimated that another 80,000 who observed the game on TV or radio were, like myself, avid fans. Each of these 100,000 fans had some concern with the outcome of the game, and each had about 5 REM periods that night. Let's assume that, out of 500,000 REM periods, only 1 percent of the dreams (5000) involved a specific score about the upcoming game that the dreamer remembered upon awakening. There are about 36 likely score combinations in a Stanley Cup final game (each team scoring 0 to 6 goals, no ties allowed). Thus, the probabilities are that 139 dreamers (5000 divided by 36) predicted the correct score. Since there are about 36 players on the two teams, roughly the same odds exist that the player who scored the game-winning goal would be identified by 139 dreamers. Picking both the final score and the goal scorer, or picking either, is a fairly unlikely event for any individual—139 out of 100,000, or roughly one-tenth of 1 percent. But the odds are clearly in favor of at least one or more individuals picking the correct score. It's comparable to winning a lottery: Your own chances of winning may be a million to one, but it is a certainty that someone will win. My theory is bolstered by the recent case of a woman who twice won the million dollar lottery in New York.

Probability theory can explain the outcome of dream telepathy experiments. While most laboratory studies have produced negative results, a few researchers remain convinced on the basis of interesting but meager findings that dream telepathy does occur. In these studies, observers attempt to influence or guess the content of a sleeping volunteer's dreams. The few positive results may merely represent those chance occurrences that result from carrying out hundreds of such telepathy experiments.

## Interpreting Dreams

Throughout history, humans have tried to understand the function and meaning of dreams. Eight thousand years ago Assyrian cuneiform dream books detailed the content of dreams and their meaning. Assyrians' dreams were commonly about family; about sex, teeth, animals; about hair falling out; about finding one's self naked in public—in content very like the dreams of the twentieth century. Perhaps the basic issues in life have not changed very much.

The Assyrians believed that dreams were communications from evil spirits of the deceased. The Egyptians, on the other hand, believed that dreams were a message from the gods, typically urging repentance, warning of dangers to come, or supplying answers to questions. The profession of dream interpretation, the closest analog to psychoanalysis, was well respected, and special temples were erected for its practice—all under the guidance of the Egyptian god of dreams, Serapis. At one nineteenth-century excavation site the following inscription was found over a doorway: "I interpret dreams, having the gods' mandate to do so. Good luck if you enter here." Not that much has changed. Dream interpretation, even today, is still more of an art than a science.

The notion that God will enter our lives via dreams to provide solutions to our problems was passed on to the Greco-Roman civilization by the Egyptians. (Aristotle—who based his theories on observation rather than tradition—having observed that animals dreamed, rejected the premise that dreams were supernatural.) In the Middle Ages, the formula was slightly altered; dreams were thought to be brought on by evil spirits. Martin Luther, the Reformation leader, believed that "filthy dreams" could further self-improvement by showing the dreamers their sinful ways. Eventually, however, Luther found these dreams so disturbing that he prayed for God to communicate with him some other way.

Sigmund Freud was one of the first investigators to make the study of dreams a legitimate area of medical inquiry. Freud's data was derived primarily from the dream reports of neurotic and borderline-psychotic patients. Although his work, *The Interpretation of Dreams*, sold few copies when it was first published in 1905, it soon became the basis for the dominant dream theory of the twentieth century. Since sleep was conceived of as a passive or "off" state by Freud and other physicians, dreams were thought to be the guardians of sleep, protecting the sleeper from being awakened by external stimuli. Instead of interrupting sleep, for example, noise would be incorporated into dreams as an image, thus preserving sleep. More importantly, dreams were the royal road to the unconscious, a series of disturbing sexual and aggressive wishes and impulses that were disguised to various degrees. Nightmares occurred, the Freudian theory held, when the dreamer's impulses broke through the disguises or defenses. But the meaning of most dreams could be discerned only by analyzing dream symbols and their associations. This would enable the dreamer to uncover and become aware of formerly forbidden impulses.

## Dream Symbols

If you remember a dream, what is the best way to decipher its meaning? The answer depends in part upon who is interpreting the dream. It has been said by psychotherapists that patients in Freudian therapy dream about sexual themes, while Jungian patients dream about mysticism—and perhaps a patient being treated by a behavioral therapist would be taught not to remember any dreams at all! Let's take a look at two dreams with the same symbol—a hat—and see what Jung and Freud do with it.

### Dream Reported by Jung's Patient

The dreamer is at a social gathering. On leaving, he puts on a stranger's hat instead of his own.

*Jung's Interpretation*

> The hat, as a covering for the head, has the general sense of something that epitomizes the head. Just as in summing up we bring ideas "under one head" [*unter einen Hut*], so the hat, as a sort of leading idea, covers the whole personality and imparts its own significance to it. Coronation endows the ruler with the divine nature of the sun, the doctor's hood bestows the dignity of a scholar, and a stranger's hat imparts a strange personality. Meyrink uses this theme in his novel *The Golem*, where the hero puts on the hat of Athanasius Pernath and, as a result, becomes involved in a strange experience. It is clear enough in *The Golem* that it is the unconscious which entangles the hero in fantastic adventures. Let us stress at once the significance of the *Golem* parallel and assume that the hat in the dream is the hat of an Athanasius, an immortal, a being beyond time, the universal and everlasting man as distinct from the ephemeral and "accidental" mortal man. Encircling the head, the hat is round like the sun-disc of a crown and therefore contains the first allusion to the mandala [a ritualistic or magical

circle]. . . . As a general result of the exchange of hats we may expect a development similar to that in *The Golem*: an emergence of the unconscious. The unconscious with its figures is already standing like a shadow behind the dreamer and pushing its way into consciousness.[16]

## Dream Reported by Freud's Patient

"I am walking in the street in summer; I am wearing a straw hat of peculiar shape, the middle piece of which is bent upwards, while the side pieces hang downwards (here the description hesitates), and in such a fashion that one hangs lower than the other. I am cheerful and in a confident mood, and as I pass a number of young officers I think to myself: You can't do anything to me."

*Freud's Interpretation*

As she could produce no associations to the hat, I said to her: "The hat is really a male genital organ, with its raised middle piece and the two downward-hanging pieces." It is perhaps peculiar that her hat should be supposed to be a man, but after all, one says: *Unter die Haube Kommen* [to get under the cap] when we mean "to get married." I intentionally refrained from interpreting the details concerning the unequal dependence of the two side pieces, although the determination of just such details must point the way to the interpretation. I went on to say that if, therefore, she had a husband with such splendid genitals she would not have to fear the officers; that is, she would have nothing to wish from them, for it was essentially her temptation phantasies which prevented her from going about unprotected and unaccompanied. This last explanation of her anxiety I had already been able to give repeatedly on the basis of other material.

It is quite remarkable how the dreamer behaved after this interpretation. She withdrew her description of the hat, and would not admit that she had said that the two side pieces were hanging down. I was, however, too sure of what I had heard to allow myself to be misled, and so I insisted that she did say it. She was quiet for a while, and then found the courage to ask why it was that one of her husband's testicles was lower than the other, and whether it was the same with all men. With this the peculiar detail of the hat was explained, and the whole interpretation was accepted by her.

The hat symbol was familiar to me long before the patient related this dream. From other but less transparent cases I believed that I might assume the hat could also stand for the female genitals.[17]

Most psychotherapies that postulate the existence of the unconscious use dream interpretation. Most therapists do not use a dream deciphering code. Instead, patients are asked to discuss and associate to their dreams. For one dreamer a hat could be a sexual object, for another it might represent a defense or a way to hide one's thoughts. In that manner each dreamer develops his or her own symbolic understanding. According to both Jung and Freud, however, there may be a few universal symbols that mean the same for each person. For Freud, dreams about teeth falling out concern masturbation; dreams of falling are about sexual anxiety; dreams of swimming represent the repressed childhood pleasure associated with bed-wetting. For

*"All right, have it your way—you heard a seal bark"*

Jung, phallic symbols in dreams represent fertility, power, or healing, rather than sexual themes. Animals such as the horse in a dream represent the nonhuman aspect of our personality, such as instincts or hidden feelings of power.

### Dreams and REM Sleep

Dreaming and REM sleep are controlled by the oldest and most primitive part of our brain, the brain stem—the same brain area that controls basic regulatory functions such as respiration and temperature regulation. This brain area is easily recognized in mammalians, all of whom show evidence of REM sleep. There is evidence that animals, as well as humans, probably have visual imagery or dreams during their REM periods. In 1964 Dr. C.J. Vaughan trained rhesus monkeys—in a waking state—to press a bar whenever an image flashed on a screen; they continued to do this in REM sleep but not in NREM sleep.

With the exciting discovery in the 1950s that rapid eye movement (REM) sleep was associated with dream recall, researchers initially hoped the mysteries of dreaming would be fully unraveled. Studies were quickly developed to test Freud's theories. Subjects were sprayed with water and subjected to noises and to warm and cold room temperatures to see if these stimuli would be incorporated into their dreams. Occasionally, when the dreamer would awaken, the stimuli did indeed appear to be incorporated: "I was standing under a waterfall," or "I dreamed of an explosion." However, in most cases the dreams were rather mundane, without evidence of Freud's sleep incorporation phenomenon. Furthermore, even when dreamers could be awakened during the most intense periods of rapid eye movement, the dreams recovered were usually not about sex or aggression. In fact dreams were often about the experience of being studied overnight in a sleep laboratory. The more sleep researchers tried to validate Freudian dream theory the more apparent it became that Freudian dream theory rests on the existence of the unconscious and on the analyst's ability to help the patient decode dream symbols. It is very hard to measure the unconscious. Sleep research methods have therefore not really fulfilled the promise of validating the Freudian approach.

Other psychiatric dream theories were evaluated. Individuals selectively deprived of REM sleep were at first thought to become psychotic because their unconscious impulses could not be properly discharged, but controlled studies revealed that this is not the case. Those deprived of REM sleep will try to increase the amount of REM sleep the following night—the so-called REM rebound—but psychological or physiological ill effects are limited. Subjects have been deprived of REM sleep for up to sixteen nights without significant effect. It becomes increasingly difficult to continue the deprivation over a succession of nights because, upon being put to sleep, deprived subjects enter REM sleep almost immediately.

The postulate that insanity was equivalent to a waking dream, first advanced by the neurologist Hughling Jackson, was disproved. The hallucinations of schizophrenics during wakefulness are not related to the dreams of normals. In fact, schizophrenics have fairly normal REM cycles, although their dream content evidences a more acute anxiety and sense of emerging and impending doom than does that of normal persons.

REM sleep and dreaming may serve as an adaptational and emotional process. Among individuals subjected to anxiety-provoking pre-sleep stimuli, such as disturbing, violent movies, those who dream about these experiences are less anxious and tense the following day than those who do not. The amount of REM sleep of recently divorced women has been shown, in recent studies, to increase, supporting the concept that dreaming serves to enhance the capacity to cope with emotional problems. Another interesting adaptive theory is the "sentinel hypothesis": Dreaming is a periodic arousal-security system that allows the sleeper to scan the environment for any threatening signs. If everything is safe, sleep continues and external noises will be incorporated in dreams. If the environment is not safe, the sleeper will awake.

Over the past thirty years, since the discovery of REM sleep, researchers have collected a series of important facts about this unique stage of sleep:

**1.** REM sleep occurs in mammals and to some extent in birds.

**2.** REM sleep is controlled by the primitive brain stem.

**3.** REM sleep is more prevalent in infants, occupying up to 50 percent of their sleep time, but declines to 20 percent in adult years.

**4.** After REM sleep deprivation, REM rebound occurs but is usually not associated with significant psychological effects.

**5.** The intrusion of REM sleep into wakefulness (narcolepsy) is maladaptive.

**6.** Approximately 90 percent of individuals awakened during REM sleep will report a vivid dream.

**7.** Many remembered dreams seem to be psychologically meaningful to the dreamer, but repetitive dream reports collected in the laboratory suggest that the content is often mundane.

**8.** The timing of REM sleep is precise, and different clusters of neurons switch on and off with the start and ending of each REM period.

**9.** There is no obvious correlation between the amount of REM sleep and type of animal species or brain size of animal.

### The Dream Generator

This new information has inspired several attempts to develop a comprehensive theory of dreaming, that is, to pursue answers to the large questions, What are dreams? and What function do they serve? The *activation-synthesis* theory of dreaming asserts that dreaming is merely the incidental outcome of an essential biological process. Researchers have demonstrated that during REM sleep a group of nerve cells in the brain stem—gigantocellular tegmental field (GTF) neurons—fire; they constitute the "on switch." The excitation of the cells is normally inhibited by another group of neurons in the brain stem—the locus cerules (LC) cells, or "off button." These cells fire in NREM sleep, through a biochemical process as yet not well understood.

As REM sleep begins, the activity of the LC neurons diminishes and the GTF cells become excited, reaching a peak during bursts of rapid eye movements. The GTF cells have connections (pathways) to the visual and sensory brain centers as well as to parts of the brain where emotions originate. Thus, the brain stem, or lower brain, bombards the cortex with a series of random impulses, which the higher brain tries to weave into a coherent pattern of experience. This theory contrasts sharply with the Freudian model, which pinpoints psychological meaning as the main purpose of dreaming. In the activation-synthesis model, the primary purpose of dreaming, and therefore of REM sleep, may be to serve as a test system for the brain ("brain development and maintenance") in order to enhance thinking during wakefulness. In any case, dreaming is viewed as a purely physiological process and any psychological meaningfulness is merely fortuitous. To repeat the thought with which this chapter opens, "A dream about a banana may be a dream about a banana."

Many contemporary dream theorists have followed similar tracks. Professor Francis Crick, Nobel laureate for his research on DNA, published his theory of dreaming in the journal *Nature* in 1983. Professor Crick suggested that it might actually be harmful to remember dreams! REM sleep and its accompanying dreams, he conjectured, are necessary; they enable the brain to discard unwanted or unsuccessful modes of brain behavior, such as fantasies, useless information, and superfluous or disturbing associations with previous wakefulness. Thus the function of dreaming is to provide a reverse learning at night.

Inevitably, the computer has influenced dream theory by providing a useful metaphor: If you use a personal computer, you are apt to find that after a time you have accumulated numerous data files, and that you no longer remember the content of some of them. When the computer starts to run out of storage space, you need to review old files, erasing the information you no longer need and refiling and reorganizing data that you want to save. You would probably do this in the evening or on the weekend, when normal computer functions are down. Dreams may be the remembered fragments of the brain's nocturnal data processing. The off-line brain assimilates the day's experiences and updates the old "programs." This theory is interesting because it suggests an explanation for why we need to be asleep— to move into off-line mode—and why we dream. It would be difficult to update our memory bank while we are awake and functioning.

Dreaming must serve some important purpose; if it doesn't, why is the REM system so prevalent and precise? My own view is that dreams are initially a series of random images, and that the way the dreamer relates to them invests them with psychological meaning. In dream analysis, the task is to distinguish the signal from the noise. Dreaming may function as a way for the brain to retrain itself, to analyze some information, and to discard other information at a time when it is safe from interruption by the distractions of the waking world. Maybe dreams are bits of information that the nocturnal computer has selected for special editing, either to be understood or discarded. There is so much information to digest today that we should be thankful the brain is active, that it is doing something for us while we rest and take a respite from incoming data.

## Creative and Lucid Dreaming

One of the most fascinating things we know about dreams is that several major scientific discoveries have been revealed in dreams. The structure of the benzene ring, the chemical transmission of neural impulses, the invention of the sewing machine—all were riddles solved during dreaming. A remarkable dream led to the deciphering of the cuneiform tablet called the Stone of Nebuchadnezzar. In 1893, Professor Herman Hilprecht, an archeologist who had been searching for the meaning of this tablet, had the following dream about two small dig fragments that were housed in separate cases in an Istanbul museum: "A tall, thin priest proclaimed that these two tablets belonged together." Hilprecht went to the museum and found that the two pieces fit together to form a meaningful inscription, allowing him to decipher the Stone of Nebuchadnezzar.

Several dream researchers have advocated techniques to improve dream recall and so make better use of those nocturnal messages. One school advocates increasing *lucid dreams*—the experience of being aware during the dream that you are dreaming. Through the use of auto-suggestion just prior to falling asleep, it is possible to increase the number of lucid dreams. If you are plagued by monsters in your dreams, the next time you encounter the monster you can tell yourself in the dream to reach out and shake hands and make friends with the monster. This may lead to your discovering the nature of the anxiety in the dream and lead to your dealing more effectively with unacceptable parts (monsters) of your own personality.

Marquis d'Hervey de Saint-Denys, a nineteenth-century professor of Chinese literature by day and a dream researcher by night, described his technique for overcoming a dreadful recurrent nightmare.

> I was not aware that I was dreaming, and imagined I was being pursued by abominable monsters. I was fleeing through an endless series of rooms. I had difficulty in opening the doors that divided them, and no sooner had I closed each door behind me than I heard it opened again by the hideous procession of monsters. They were uttering horrible cries as they tried to catch me. I felt they were

gaining on me. I awoke with a start, panting and bathed in sweat. . . . On the fourth occurrence of the nightmare . . . just as the monsters were about to start pursuing me again, I suddenly became aware of my true situation. My desire to rid myself of these illusory terrors gave me the strength to overcome my fear. I did not flee, but instead, making a great effort of will, I put my back up against the wall, and determined to look the phantom monsters full in the face. This time I would make a deliberate study of them, and not just glance at them, as I had on previous occasions. I stared at my principal assailant. He bore some resemblance to one of those bristling and grimacing demons which are sculptured on cathedral porches. Academic curiosity soon overcame all my other emotions. I saw the fantastic monster halt a few paces from me, hissing and leaping about. Once I had mastered my fear his actions appeared merely burlesque. I noticed the claws on one of his hands, or paws, I should say. There were seven in all, each very precisely delineated. The monster's features were all precise and realistic: hair and eyebrows, what looked like a wound on his shoulder, and many other details. In fact, I would class this as one of the clearest images I had had in dreams. Perhaps this image was based on a memory of some Gothic bas-relief. Whether this was so or not, my imagination was certainly responsible for the movement and colour in the image. The result of concentrating my attention on this figure was that all his acolytes vanished, as if by magic. Soon the leading monster also began to slow down, lose precision, and take on a downy appearance. He finally changed into a sort of floating hide, which resembled the faded costumes used as street-signs by fancy-dress shops at carnival-time. Some unremarkable scenes followed, and finally I woke up.[18]

And his nightmares ended.


### Dream Sequences

REM periods recur on a regular basis every 90 minutes during sleep, suggesting that REM sleep can best be viewed as an ultradian biological rhythm. Even though you have about five REM periods a night, each lasting anywhere from 15 to 45 minutes, your dreams do not repeat. It's rare for a subject in the laboratory to report the same dream in awakening from different REM periods.

There are, however, occasions when the same theme appears to be repeated in five different dream awakenings through the night.

**1.** The dreamer is in a swimming pool, with everyone admiring his body. He sees a possible competitor for the attentions of a friend (sex undetermined), but "for some reason I wasn't as afraid as I ought to have been. I seemed to have some sort of metamorphosis. I became just like a strong-muscled Greek god." He excels in a series of diving demonstrations, once almost losing his bathing suit in the process. The dream ends with the subject waiting to watch a TV program featuring a prominent Hollywood actress.

**2.** The subject is in a room with two Hollywood entertainers, one of whom fires his gun at the door causing it to collapse. A half-visible man walks in, demanding the "plans." He walks over to the gun-shooting entertainer and begins to choke him. But the dreamer attacks the assailant, and knocks "the hell out of him. . . . I remember standing there in kind of a triumph."

**3.** The dreamer is trying to enter a room by way of the window, because he lacks a key to the door. An acquaintance standing by the door gives the subject two sandwiches, the "worst ones imaginable." The two go in, but the subject is not satisfied with what he sees and wants to leave. "And there was something about nitroglycerine. . . . The last thing was somebody throwing a baseball."

**4.** The dreamer is making inquiries concerning the underground movement during World War II. At dinner time, he asks his mother about it and also asks her, "Don't you think we can settle the question?"

**5.** The dreamer watches an argument erupt during a classroom economics lecture. The lecturer continues as two male students argue about the danger of inflation.[19]

In each of these dreams there is a recurrent theme of conflicts among males. The solutions to the conflicts range from magic (transformation into a Greek god) to aggression (shooting, war, nitroglycerin) or possibly negotiation (argument, discussion). Although the dreams are not identical there is a common thread. These thematic connections are easier to see in hindsight. Experimenters trying to sort out a stack of fifty dream reports by ten dreamers have great difficulty in matching each set of five dreams to the correct individual.

## REM-Related Sleep Disorders

### Nightmares

The most common disruption of REM sleep is the *nightmare*—vivid anxiety-provoking dreaming frequently associated with awakening. As they most commonly start in childhood, nightmares are generally believed to be related to the psychological conflicts of childhood. Nightmares in adults, however, are a common response to real or threatened traumatic events—rape, murder, war, automobile accidents—all events that threaten our survival and therefore our sense of security.

Nightmares can also be induced in a psychologically healthy person by physiological manipulations. Any substance that chemically reduces REM sleep (alcohol, some kinds of sleeping pills, sleep deprivation) will eventually be followed by REM rebound. This is frequently experienced as intense, vivid dreaming, "too much dreaming," or nightmares. Chronic alcoholics experiencing delirium tremens during acute alcohol withdrawal may have 100 percent REM sleep, perceived as fitful, hallucinatory, and frightening sleep.

Nightmare sufferers can often be helped with psychotherapy or controlled withdrawal from REM suppressant substances. A rarer and more bizarre disorder, *night terrors*—from which patients abruptly wake up in a panic within the first hour of NREM sleep—may not be readily treatable. Night terror victims may be only partially coherent after the episode, indicating that they are neither fully asleep nor awake, and typically by morning do not remember the episode. Most night terrors are associated with rapid heart rate and respiration and feelings of tension and panic. Often little content is remembered upon the nocturnal awakening except for a terrifying fragment or emotion: "I was scared of being attacked," "I was suffocating," "The room was closing in." The vivid, detailed content associated with REM dreams is absent.

Most night terror sufferers outgrow these attacks; in general they disappear after childhood and adolescence, suggesting that night terror may be related to conflicts of psychological identity. On the other hand, certain drugs that reduce the arousal threshold, making it more difficult to awaken, can suppress night terrors, suggesting it is a disorder of arousal: Patients may wake up because their body is in an energized state and consequently react with a feeling of panic and terror. The thought fragments remembered may be the waking brain's method of explaining the sudden intense physiological arousal. In general, night terrors in children are best left untreated; getting the child quickly back to sleep is most effective since it is all forgotten in the morning. In adults, night terror usually warrants psychological intervention.

Have you ever awakened from sleep and been temporarily unable to move? *Sleep paralysis* is a frightening phenomenon that most people have experienced at least once. Repeated attacks of sleep paralysis are very rare except in patients with narcolepsy, or in a few case reports of families in which all members are affected, suggesting a genetic basis. Usually sleep paralysis is simply a slight deviation from the normal REM sleep-to-wakefulness transition. During REM sleep our reflex system is paralyzed, preventing us from acting out our dreams. There is an excellent chance that the final morning awakening will occur during REM sleep. If we wake up rapidly (perhaps due to a disturbing dream or noise in the environment) we may feel that our body is paralyzed, i.e., still in REM sleep, but our consciousness is returning—we are entering a state of wakefulness. The consequence of being abruptly awakened from a hallucinatory experience (dreaming) but unable to move is a feeling of terror, a fear of losing control. Most people try to fight sleep paralysis. They struggle to wake up, but this usually prolongs the frightening episode. A few patients have told me that if they simply try to sleep rather than trying to wake up, the episode ends and they later awaken without difficulty.

## Impotence

At the Baylor University Sleep Center in Texas, over 60 percent of all clinic patients are referred because of impaired sexual potency. The observation that 95 percent of REM periods are associated with full or partial penile erections has led to a novel method for the diagnosis of the cause of impotence. By monitoring erection cycles during REM sleep, it is possible to determine the cause of impotence. Patients who complain of impotence but demonstrate normal nocturnal REM erections are diagnosed as psychologically disturbed, whereas those without REM erections are presumed to be organically impaired. Diagnosing impotence properly is important because of the treatment possibilities: Patients with primarily psychologically based impotence will benefit from psychotherapy or sexual counseling, whereas patients with organically based impotence may receive hormone therapy or surgical implantation.

There is currently a great deal of controversy over the question of how best to determine whether nocturnal erections occur. Sleep clinicians advocate a simultaneous sleep and erectile study. In this manner the clinician is certain that a normal REM sleep has occurred. Many urologists and "potency clinics" rely on a stamp test. A roll of postage stamps is placed around the penis at bedtime. If the roll of stamps is broken, this would indicate normal nocturnal erections.

The prevalence of impotence, or erective impairment, increases with age, from 2 percent for men less than forty years old to 75 percent at age eighty. Whereas the Kinsey report suggested that diminished physiological capacity was the cause, Masters and Johnson suggested that certain patterns of sexual activity, or inactivity, were the cause. Careful all-night studies as well as daytime urological testing suggest that 70 percent of patients complaining of chronic impotence have an organic basis for their disorder, such as diabetes, side effects from medications, or vascular diseases of the genitals or nervous system. (Many physicians do not actually examine the penis although certain vascular diseases can be easily diagnosed from physical inspection.) Nocturnal penile tumescence (NPT) monitoring has now expanded beyond Texas, with dozens of centers spread around the United States performing such diagnostic evaluations.

### REM Sleep and Your Heart

Nocturnal heart attacks in chronic cardiac patients, presumably during sleep, are fairly common. These incidents may be related to REM sleep, when heart rate is most variable. More recently, frequent reports of normal healthy young males, Southeast Asian immigrants to the U.S., dying during sleep has attracted the attention of sleep researchers. This disorder was already recognized throughout Asia as *bangungat*—"to arise and moan" (Phillipines); *non-laitai* "sleep death" (Laos); and *pokkuni* "sudden death" (Japan). All-night cardiovascular monitoring in three survivors of these attacks revealed ventricular tachycardia—dangerous, rapid arrhythmic increase in heart rate during sleep.

Christian Guilleminault at the Stanford Sleep Disorder Clinic discovered that other patients show cardiac abnormalities only during REM sleep. One of his patients was a 30-year-old female physician and jogger. Dr. Guilleminault describes her symptoms: "Over the past two and one-half years, on the days after night duty, she had intermittent palpitations, vague chest pains that occasionally radiated to the shoulder, and lightheadedness. Sometimes, when awakened abruptly in the early morning by an alarm, she had a feeling of faintness and blurred vision without complete syncope. She sought medical attention after suddenly losing consciousness while trying to answer the telephone in the middle of the night."

Several daytime electrocardiograms, stress tests, angiograms, and catheterizations were all normal. However, 24-hour-round-the-clock Holter electrocardiograms revealed asystoles (complete cessation of heartbeat) lasting anywhere from 2½ to 9 seconds. All-night sleep studies revealed that all asystoles occurred during REM sleep; none occurred during NREM sleep. Furthermore, nearly all the cardiac pauses occurred simultaneously with a burst of rapid eye movements and were not associated with awakenings or arousals. The association with bursts of eye movement does not mean that bad dreams caused her symptoms, but rather that her disorder occurred during the time of intense physiological activation.[20]

Although cardiac dysfunction during sleep is a frequent concomitant of sleep apnea, the demonstration of cardiac arrest in young adults with otherwise normal cardiac functioning is surprising. A specific abnormality of the regulation of heart rate during REM sleep may explain some incidents of cardiac arrest during sleep. More importantly, this case study demonstrated that during REM sleep heart rate is regulated by a physiological mechanism different from that operating during wakefulness and NREM sleep. Such patients, with a disorder of this REM-cardiac system, when properly diagnosed with 24-hour electrocardiograms and sleep studies, can be treated successfully with cardiac pacemakers.

## Nocturnal Violence

Over the years, I have had several patients who engaged in violent nocturnal episodes during sleep. One patient actually walked through a window to escape the claustrophobic panic he was experiencing. Another repeatedly and unknowingly punched his woman friend during sleep. She came to the sleep clinic and I recommended couple therapy. They came for one session but the man refused to explore any psychological problem that might be responsible for his behavior.

One day a lawyer called at the sleep clinic regarding his client, a teen-age boy who, awakened by a friend about 45 minutes after going to bed, became an accomplice to a murder. His client shouted "kill him" and looked on while his friend—also a teenager—stabbed a man to death. The lawyer wanted to know whether his client had grounds for a diminished capacity defense; that is, could he claim that his actions were not under his control because of his altered state of consciousness. Although is possible for someone to be disoriented upon awakening from sleep, this effect is brief. Judgment is not substantially affected upon awakening except perhaps in a chronic sleepwalker. I later learned that he was convicted of voluntary manslaughter, which was a slightly lesser charge than the District Attorney was calling for.

In another case, a D.A. called me regarding a child molestation case. The defendant claimed he was unaware of the events since they happened during sleep. I pointed out that this seemed unlikely, since the abuse lasted over thirty minutes. In general, nocturnal violence and prolonged confusion are rare and more likely to be associated with awakenings from Stage 3-4 sleep than from REM sleep. Most behaviors during the confused transition from sleep to wakefulness are rather simple and short-lived. One would be unlikely to carry out a complex pattern of violent behavior without gaining full consciousness.

But there are exceptions. Professor Ian Oswald of Edinburgh University in Scotland recently reported the case of a man waking up from a night terror and actually killing his wife because he thought he was being attacked. He was acquitted because the jury believed that he was not fully conscious and therefore not responsible. Professor Oswald stated that precedence for this legal defense can be found in English law dating back to the seventeenth century.

### Dream Content

The number of sleep researchers studying dream content has progressively declined over the three decades of modern sleep research. During my training at the University of Chicago in 1973, there was widespread optimism that the mysteries of dreaming would be revealed. We hoped to gather data by observing dream awakenings. After midnight, my colleague Don

Bliwise and I would have put one or two volunteers to sleep, placed our take-out order at Ribs and Bibs, and settled in to wait for a "good" REM period. Don would sit in the control room watching the polygraph and would awaken the subject during either a period of intense eye movement or a quiet period. When Don found a good pattern he would buzz the subject to awaken him, and I would start a dream interview. My task was to rank the dream according to its degree of bizarreness. Our goal was to find a brain pattern that might indicate when dreams became bizarre and thus, we hoped, provide a link to daytime hallucinations of schizophrenics.

The greatest hurdle was in detecting whether it was only the language in which the subject reported the dream that was bizarre, or whether the dream itself was bizarre. Unfortunately, our results were negative, and now both of us, like so many sleep researchers, are concentrating more on sleep pathology. Until sleep researchers develop a methodology to directly assess dream content without awakening the sleeper for his verbal report this field will continue to be muddled.

No one, not even the analyst thoroughly familiar with the patient's experiences, emotions, and concerns, has been able to accurately predict the content of a dream before the onset of sleep. As we have suggested, dreams are complex, not simply related to daytime activities. Some rough guidelines exist: Dreams are more likely to involve people than animals, males than females, hostility than friendliness, unpleasantness than pleasantness, and passivity than action; color is reported in about 50 percent of dreams. There may be a thematic quality to a night's dreaming whereby a central conflict in one REM period is played out in later REM periods.

Dreaming remains something of an enigma. The vast majority of dreams are never remembered, yet humans spend about 10 percent of their lives dreaming. Is a dream that is not explained like a letter that has not been read, as the Talmud suggests? Or are most dreams junk mail? Research on dreaming has slowed down since the concentrated research and discovery of the 1950s and 1960s. In order to advance the field to a new level, new technologies will be needed. Perhaps in the future it will be possible to monitor a volunteer with electrodes and to project the brain signals onto a television screen. Then we can find out how dreams are formed, how they develop, and what happens to them after they are completed.

# 7

# "I Can't Sleep!"

*Don't go to sleep—so many people die there.*

—Mark Twain

Mark Twain's warning undoubtedly evoked laughter in his audience, but for some, behind that laughter lurked a real fear. Sleep has been called "death's younger brother," and for many insomniacs sleep is linked to a subconscious fear of death. Certainly, for some, going to sleep rekindles a fear of death; for others, it represents a fear of losing control.

One way to judge the degree of this fear is to observe pre-sleep rituals. Normal sleepers typically follow a simple routine—perhaps brushing their teeth, securing their door locks, and turning down the heat. Insomniacs who fear death or loss of control may have exaggerated bedtime rituals—checking and rechecking to make sure the stove is off, arranging and rearranging the bedsheets and pillows until everything is perfect. One of my patients counted his rosary beads every night while he prayed; the next morning he could tell how long it had taken him to fall asleep by seeing at what bead he had left off.

Some of my insomniac patients report dreams that tip me off to their fear of death: "I'm in a tight spot and I can't get out of it." "My heart's on the table and they're working on me." "Fleas are jumping around; it feels like they're biting me." These are the dream reports of a 68-year-old retired executive who delights in racing his sports cars at speeds over 100 miles per hour but is incapacitated by insomnia. Ever since his coronary bypass surgery four years ago his sleep has been uneven and interrupted.

It is difficult for normal sleepers to appreciate the sufferings of insomniacs unless they too run into a series of never-ending bad nights. Luckily, for most people insomnia lasts only a night or two, usually in association with an acutely stressful event—perhaps an important exam or an upcoming meeting or trip. While all humans probably experience some degree of insomnia at a particular point in their lives, it is extremely rare to find an animal with insomnia. Although their nervous system and sleep physiology are similar to ours, animals don't share our psyches—they are closer to nature. The bear sleeps all winter; the owl is awake all night. If cats or leopards or foxes are awake at night, it is probably because darkness enables them to do something they cannot do in the daylight—to prowl and seize other animals. Insomnia is a *human* disorder.

Inevitably linked to human activities during wakefulness, insomnia must be regarded as a 24-hour problem rather than merely a sleep disorder. The chronic insomniac—the typical patient treated in a sleep clinic—may report not having had a decent night's sleep for twenty years but is less concerned with the effects on daytime functions. Nocturnal complaints include difficulty in falling asleep and staying asleep, waking up too early, and poor quality sleep. Corresponding daytime complaints typically include fatigue, sleepiness, poor performance, aches, and anxiety. Any combination of these complaints can be found in a given patient. The role of the sleep clinic is to evaluate the complaints and to determine whether they can be verified by physiological measurement. Following that assessment, a diagnosis and a specific treatment plan can usually be provided to benefit the insomniac.

To better define insomnia complaints, Dr. Mary Carskadon, one of Stanford's most dedicated researchers, decided to observe a large group of typical insomniacs. An advertisement in a Bay Area newspaper brought in 122 self-reported insomniacs who volunteered to spend the night at the Stanford Sleep Center so that researchers could objectively measure their sleep patterns. On average, these insomniacs reported in the morning that it had taken them 60 minutes to fall asleep. The objective sleep tracings showed something quite different: The average subject fell asleep within 15 minutes of lights out. Another example of the discrepancy between the insomniacs' perception and the reality was their report of an average of only 4½ total hours of sleep, while the sleep recordings indicated that total sleep time averaged 6½ hours!

Why is there such a large discrepancy between the subjective reports of these insomniacs and the objective findings? The usual explanation is that the patients are neurotic, that they are exaggerating their complaints. This explanation is true for only a small minority—about 10 percent. There are other good reasons: Undetected, repetitive arousals during sleep can cause a sensation of little or no sleep. In 1969, when the first patients in the United States to be treated for sleep disorders came to the Stanford Sleep Clinic, we did not routinely put muscle electrodes on their legs to detect abnormal leg movements, or thermistors near the nose and mouth to detect irregular breathing patterns. At that time, up to 50 percent of patients complaining of insomnia were diagnosed as pseudo-insomniacs—that is, they complained of insomnia but exhibited no objective evidence of impaired sleep. With increasingly sophisticated technology and the discovery of new sleep pathologies, objective findings are now detected in many of the "neurotic" insomniacs.

### Insomnia: Who Has It?

Insomnia is believed to be a problem for 15 to 30 percent of the adult population. In some special groups—shiftworkers or psychiatric patients—the prevalence is closer to 65 percent. Prevalence rates are fairly similar in various countries where insomnia has been studied—England, Finland, Israel, and Italy.

In order to distinguish insomnia patients, studies comparing good and poor sleepers have been conducted in the laboratory as well as in large populations. One of the earliest sleep laboratory studies found that poor sleepers were more anxious and more physiologically aroused than good sleepers. In general, however, the brain-wave patterns of insomniacs are very much like those of normal sleepers. So as far as we know, insomnia does not seem to be a result of abnormalities of the brain.

One of the most thorough epidemiological studies of sleep disorders was done in San Marino, an independent republic of 20,000 inhabitants located in central Italy. All of the citizens participate in a free national health service and were therefore available for questioning. Twenty-five percent of the population (a representative cross-section of the republic) completed a sleep questionnaire. The results showed that 13.4 percent of San Marino's inhabitants "always or almost always slept badly without using sleeping pills." Good sleepers averaged 9 to 10 hours of total sleep time if under age fifteen, 7 to 8 hours if over age fifteen, and typically fell asleep within 10 minutes. The young did not recall waking up; the older subjects reported waking up once. Good sleepers rarely recalled dreams or took naps, and daytime sleepiness, if present at all, was usually confined to the commonly reported mid-afternoon dip.

Poor sleepers of any age, on the other hand, typically took 30 minutes to fall asleep, woke up several times during the night, and averaged about 2 hours less sleep time per day. They were likely to remember dreams, to nap during the day, and to feel drowsy throughout the day. Although poor sleepers used less caffeine than good sleepers, they used more sleeping pills. They were usually older and more likely to be anxious or depressed, to have family and/or work problems, and to be female.

In a similar study in Houston, Texas, the prevalence of insomnia was also higher for females than for males, for retirees than for working people, for the widowed, divorced, or separated than for the married or single, for renters than for homeowners—and inversely related to socioeconomic status and family income.

These results suggest that insomnia is closely allied to the degree of one's emotional stress and recent lifestyle changes, especially from age 20 to about age 65. After that, insomnia may be more closely related to physiological sleep disorders and medical problems. A group of studies carried out at Stanford confirmed this trend: Physiological sleep disorders such as sleep apnea and periodic movements in sleep were the most common cause of insomnia in elderly patients.

What impact does chronic insomnia have on performance and health? Surprisingly, 48 percent of respondents believe poor sleep habits can lead directly to heart disease. Although, as sleep clinicians, we usually tell our patients that chronic insufficient sleep does not lead to any specific disease (this is true since we cannot measure or detect any abnormal physiological events following sleep deprivation studies), there is some evidence for an association between good sleep and longevity (see Chapter 4). Individuals who sleep 7 to 8 hours have the highest survival rate.

As for performance, Laverne Johnson and his co-workers at the Naval Health Research Center in San Diego have determined that insomnia may be a key factor in predicting future performance of naval personnel. After six years of follow-up, sailors who were poor sleepers had fewer promotions, poorer recommendations, higher attrition, and more medical hospitalization than good sleepers. These results suggest two hypotheses: (1) poor sleep causes drowsiness, poor health, and hence poor performance, or (2) the complaint of insomnia indicates a person with chronic psychological problems that extend beyond sleep.

Clinical experience supports both these hypotheses. Chronic insomnia patients have been most thoroughly evaluated at the Association of Sleep Disorder Clinics (ASDC) across the United States. The evaluations are based on examinations by physicians, psychologists, clinical polysomnographers (the current term for sleep specialists), and on psychological tests, blood tests, and nighttime sleep studies. These patients represent a group of severe chronic insomniacs. The results of a complete evaluation of nearly 8000 clinic patients support the conventional wisdom that insomnia is caused by depression or stress. The most common diagnostic categories are insomnia associated with depression or other psychiatric illness (35 percent) and stress-conditioned insomnia (15 percent). Yet that still leaves 40 percent with other disorders, including drug dependency, periodic leg movements during sleep, abnormal respiration, and chronobiological disorders, and about 10 percent with no objective polysomnographic abnormalities—the old pseudo-insomniac group.

### Diagnosis and Treatment

The most important consideration in relieving insomnia is to obtain the correct diagnosis. Once you have a handle on the problem, the proper treatment is usually available. Obtaining a proper diagnosis, however, is not easy; there are over twenty subtypes of insomnia. Patients themselves can represent a hindrance. Knowing they will be granted little sympathy if they mention some of their miserable daytime symptoms—fatigue, sleepiness, aching muscles—or ask for sick leave, insomniacs tend to keep their sleep complaints to themselves. Fearful of giving themselves away, they may have to be careful not to awaken their sleeping partners. Insomniacs are seen as neurotics who should have more will power. Small wonder that they prefer to rely on medication and to go on secretly living with insomnia. To avoid facing their underlying problems, insomnia patients with significant psychological symptoms may

Doctor: *How do you sleep?*
Patient: *I can't sleep at night. I walk around all night.*
Doctor: *Oh! You're a somnambulist.*
Patient: *No, I'm a night watchman.*

hold on to their insomnia rather than follow recommendations that they pursue psychological treatment.

Another major obstacle to effective treatment is poor training of physicians. Most have had no training in basic sleep-wake physiology or pathology. Their education in these areas is often provided by pharmaceutical representatives whose expertise is limited to sleeping pills. In one observational study conducted by the Stanford group, internists were found to spend an average of two minutes with patients who brought up a sleep complaint—and this included the time to write out a prescription! Thanks to the education the ASDC has been providing to medical schools in recent years, increasing physician awareness has led to a decrease in the automatic prescribing of hypnotic drugs as well as to the realization that referral to a sleep specialist may be appropriate.

There are five major categories of *chronic* insomnia, each with its own treatment regime. *Transient* insomnia, a few bad nights of sleep due to a specific event like an exam or a crucial job interview, generally does not need diagnosis and will remit spontaneously in a few nights. A perfect example is the sleep of Olympians.

---

### The Sleep of Olympians

The skill levels of Olympic athletes from all countries are incredibly high. In many sports a split second makes the difference between a Gold Medal and oblivion. Rifle shooters are taught to squeeze the trigger in between heartbeats! You would expect that performance at this level would indicate perfect sleep. And, indeed, I found that compared with other young adults the Olympic athletes were generally excellent sleepers. Only 3 percent reported significant insomnia, and another 10 percent reported occasional difficulty in falling asleep or staying asleep. As a group, they averaged about 7¾ hours sleep a night and napped about 30 minutes a day.

The picture changed drastically, however, the night before a major athletic event. Over half reported sleep problems on that night—most commonly, difficulty in falling asleep. But early morning awakenings, poor quality sleep, and interrupted sleep were also common. In order to adjust to events scheduled early the next day, many went to bed earlier and woke up earlier than usual. For a considerable number, this meant an adjustment to earlier hours—in other words, against the natural direction of the biological clock. On average, they slept from 1 to 3 hours less than usual on the eve of their scheduled performance.

Many of the young Olympians believed that their sleep troubles had an adverse effect on their athletic performance. Specific effects were: "low energy level," "weakness," "anxiety," "inability to concentrate." Several reported that the degree to which loss of sleep affected performance was in inverse proportion to the level of competition: The stiffer the competition, the more likely they were to overcome their fatigue. But we can't always count on this burst of energy.

---

## Five Chronic Insomnias

*Psychiatric Insomnia*
The most common category (35 percent) is insomnia caused by psychiatric disorders. In these patients, objective measurements of poor sleep are closely related to a host of other symptoms, the most common being depression and anxiety. When the depression or anxiety is successfully treated, the insomnia also remits. The most appropriate treatment is a comprehensive psychiatric approach, including anti-depressant and anti-anxiety medications as well as counseling. With these patients it is more appropriate to focus on the chain of underlying causes rather than on the insomnia itself. In fact, I often make it a rule that after the first few sessions, the patient cannot talk about sleep problems. To the patient's surprise, other problems quickly surface. Why these psychiatric patients develop insomnia while others develop headaches, stomachaches, or other psychosomatic disorders is not understood.

*Stress-Conditioned Insomnia*
For about 15 percent of chronic insomniacs, stress is at the root of their problem and is a cause of intermittent insomnia. By definition, patients in this category do not have any psychiatric disorders; their symptoms fluctuate throughout the year as a reaction to life events. Their major problem is anxiety concerning whether they will sleep well and perform well the next morning. Many of these patients "try very hard" to fall asleep, but the ensuing anxiety and tension serve to arouse them further. Frequently their sleep problem persists even when the stressful life events that triggered the situation are long since past.

"I'm fed up. I can't keep going on like this!" a forty-year-old investment banker proclaimed to me. Happily married, with three young children, Mr. S's life appeared in order—no turmoil or psychiatric problems. His sleep study showed 4½ hours of uninterrupted sleep without physiological abnormalities. His major concern was whether his sleep would affect his work performance, which was highly dependent upon being alert and upon making good and accurate decisions.

He devised this treatment for himself: Drink a few glasses of wine after dinner to become drowsy. Get into bed around 8:00 P.M. to try to catch some sleep. Try really hard to stay in bed and fall asleep by counting sheep. Pray while in bed.

Unfortunately, this technique did not work. The harder he tried to fall asleep, the worse his insomnia became. Patients with a stress-conditioned insomnia can be treated rather successfully with behavioral techniques. Undoing the vicious cycle of negative conditioning and substituting a positive approach to sleeping improves most of these patients in five weeks. A paradoxical technique is used: If you stop trying so hard to go to sleep, the easier it will be to fall asleep. Mr. S was completely cured in four weeks by following the techniques suggested at the end of this chapter. However, in cases with multiple diagnoses, treatment is more difficult.

*Physiological Insomnia*
In about 30 percent of patients complaining of chronic insomnia a complete evaluation will uncover a physiological cause for their sleep disorder. Within this category are a number of interesting syndromes:

Periodic leg movements in sleep. There are patients who jerk their legs almost every 30 seconds throughout the night, often resulting in brief arousals or full-blown awakenings. A bed partner who is kicked repeatedly throughout the night may be the first one aware of the condition. Some patients with "restless leg syndrome"—unusual sensations of ants crawling in the calf or electricity in the legs delaying sleep onset—also have these periodic movements.

Sleep-related respiratory abnormalities. These patients frequently wake up anxious, with a feeling that they have not been breathing, and are unable to go back to sleep. On the other hand, many insomnia patients who actually are afflicted with apnea (spells of not breathing) complain of insufficient sleep but are totally unaware of their abnormal respirations. (Sleep-related respiratory disorders are most commonly associated with excessive sleepiness and will be discussed in Chapter 8).

Medical disorders. Another large group is composed of patients with identifiable medical disorders, such as arthritis, which cause insomnia. Biological rhythm disorders can also cause insomnia. Once a physiological sleep disorder has been accurately diagnosed, there is often a specific, effective treatment.

*Poor Sleep Habits*
This fourth category of insomnia is found both as a distinct entity and as part of other disorders. Insomnia associated with alcoholism, addiction to sleeping pills and/or other drugs, excessive caffeine use, irregu-

lar sleep schedules—all are found in this category, which accounts for 10 to 15 percent of the chronic insomniacs seen in sleep clinics.

A variety of techniques have proven helpful to these patients. However, like dieting, it takes a consistent effort and strong self-control to follow through. Most patients do better when their program is directed by a doctor, nurse, or counselor. These techniques can also be used diagnostically. If you follow all the rules for good sleep hygiene but your sleep problem persists, there is a high probability of either a psychological or physiological insomnia disorder.

*Pseudo-Insomnia—A Shrinking Category*
The final category of insomnia patient is the one whose insomnia complaints cannot be matched with any objective findings. In 1970 this group represented about 50 percent of chronic insomniacs, but thanks to refined techniques for obtaining diagnostic information, now accounts for only 10 percent of sleep patients. After a thorough all-night study, sleep tracings typically show 7 to 8 hours of normal brain-wave activity. Yet subjective estimates may be, "I had no sleep at all" or "a few hours at best." When confronted with the objective findings, patients' responses are generally either "Thank you doctor for the good news. I guess I don't have to worry about my sleep any more" or "Doctor, you must have made a mistake and scanned somebody else's sleep record. I didn't sleep at all last night."

The first group of patients make sleep clinicians feel good; the second group can be a nightmare. Some of these patients are ready to sue the doctors for malpractice; others reveal a full-blown paranoia, believing that we've actually tricked them or manipulated their brain waves. One patient from Saudi Arabia was convinced I had tricked her by exchanging sleep recordings and that I was part of a plot by the Saudi government to deport her from the U.S.! In any case, we double- and triple-check the patient for the correct sleep record and offer the best possible psychiatric referral. Several patients in the second group continue their search for confirmation of their complaints—they go on to another sleep clinic. They just can't believe they have slept all night. For these pseudo-insomniacs, denial of the facts may be a strategy to avoid facing underlying psychological problems. In some cases, our measuring techniques may not be sufficiently sensitive to detect small physiological arousals or active thinking, either or both of which may cause the perception of no sleep at all. Future discoveries are likely to reduce this diagnostic category even further.

## Specific Treatments

*Turning an Insomniac Around: Chronotherapy*

Dan J. was a 22-year-old student who came to our sleep clinic because he had difficulty falling asleep at night. He would finally fall asleep in the early morning hours but would then have difficulty waking up. This problem had plagued him since junior high school, when he was chronically late for school. At college he would frequently fall asleep at morning laboratory sessions and lectures. His social life was minimal, for which he blamed his insomnia. He had been considered lazy by his teachers and an oddball by his peers. His high-school football coach thought he lacked energy and cut him from the team. His family had developed many routines to wake him up, such as rolling him out of bed and getting him to the shower, but even there he might fall asleep again.

Dan had been to a number of doctors over the years; his unsuccessful treatments included psychoanalysis, relaxation training, biofeedback, counting sheep, sleeping pills, anti-depressant medications, and more. He tried getting into bed at a regular, early hour but this too was unsuccessful.

Our first assignment for Dan was to have him maintain a diary, a record of his sleep-wake behavior. The results indicated that on school nights he would get into bed between 10:00 and 11:00 P.M. but did not fall asleep until sometime between 3:00 and 4:00 A.M., forcing himself awake at 7:00 A.M. to get to his classes. About once a week he would sleep through his alarm and miss school. Sometimes he made it onto the bus but fell asleep, missing his stop. On weekends, however, Dan got into bed at 4:00 A.M., fell asleep within half an hour, and slept until noon. His diary showed that when he was on vacation he gave up trying to maintain a conventional bedtime. Instead, his sleep schedule was from 4:00 A.M. to sometime between noon and 2:00 P.M. During the vacation period, his complaints of insomnia and of daytime sleepiness ceased.

Dan was really a normal sleeper—on weekends and vacation. His insomnia was actually an inability to move his sleep schedule—or to adjust his biological rhythm to conventional hours. One solution would be to rearrange his social, work, and school life to conform to his sleep schedule, or to move him to a Pacific Island where the time zones were five hours behind California time. These options were not feasible. Our problem was how to move Dan's sleep schedule to earlier hours. Repeated attempts to force himself into bed at 10:00 P.M. had not worked.

After several weeks of having him submit sleep diaries we brought Dan into the sleep lab and forced him into bed at 10:00 P.M. with lights off, but he couldn't fall asleep. When we put him on the 4:00 A.M. to noon schedule, he fell asleep rapidly and slept soundly. Our laboratory studies confirmed his diary report of weekday insomnia.

At that point we decided to take advantage of two facts: (1) the clock on the wall is a circle and (2) the sleep-wake cycle naturally drifts to later hours. We decided to put Dan to bed later and later each day, in order to make him go to bed earlier! By progressively delaying his sleep time to later hours, he eventually was going to bed earlier, a technique we call *chronotherapy* (see Figure 7-1).

Over a week's time, Dan's biological clock was set to a 27-hour day. At the end of the study he was on a new sleep schedule of 10:00 P.M. to 6:00 A.M. We told Dan he was cured and could go home and stay on his new sleep schedule. His family, skeptical about his insomnia cure, had several alarm clocks set for 6:00 A.M. on his first

FIGURE 7-1

Chronotherapy: Resetting Dan's Clock

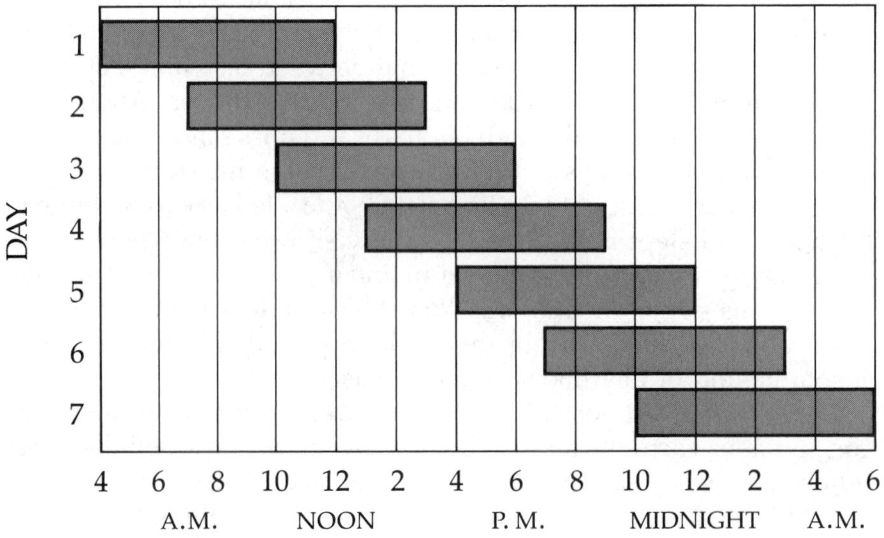

morning at home. When the alarms rang through the house, Dan's parents and brothers rushed into his room to awaken him. To their surprise he had awakened spontaneously before the alarm went off and was wide awake, eating breakfast in the kitchen. This patient, like many others with chronobiological disorders, was able to maintain improvement in a seven-year follow-up. He continued to complain, however, that although his sleep pattern was now normal, he still didn't have a girlfriend!

We have since paid more attention to the social aspects of insomnia. One self-employed psychologist who underwent successful chronotherapy decided that, after all, she enjoyed her social life better on her late-night schedule, because it allowed her to enjoy late-night activities. She therefore made the decision to drift back to her old schedule. The Hungarian playwright Ferenc Molnar was an extreme night owl and saw daylight only when he was summoned to court. Riding through Budapest to go to court he remarked in surprise at the number of people on the streets: "Are they all witnesses?"

*Electrosleep*

Electrosleep is one of a variety of treatments to improve sleep for which success has been claimed. By sending tiny electrical impulses to the brain from an oscillator, a pattern mimicking sleep is created that may actually lead to sleep. This approach had been widely heralded during the 1960s in the Soviet Union as having shown positive results in a number of patients. During a recent study tour of sleep-wake centers in the Soviet Union, which I had the privilege of leading, a Russian physician and purported sleep expert, Dr. Bull (his name has not been changed to protect his identity), a sanctioned Communist Party official, told our group that insomnia is a much more prevalent problem in the United States since it is caused by the inherent stresses of capitalism. He maintained that electrosleep is the primary effective treatment, but that in the United States pharmaceutical companies have purchased all the electrosleep devices and kept them out of circulation to better promote drugs.

With this information in mind, our group visited a sanitorium in Sochi, on the Black Sea, and asked to observe an electrosleep application. Although one of our group members volunteered to be a subject, the nursing staff had difficulty locating and operating the equipment. It had obviously not been used for some time. We never did get it going, and the clinic director, embarrassed and upset with his nurse, asked us to move on. Later on our tour, we met a bona fide Soviet sleep expert who assured us that ten years ago Soviet sleep researchers performed carefully controlled studies showing that electrosleep was ineffectual. This scientist told us that the device is infrequently used, and then mainly in the outlying provinces. In fact, there are several sleep clinics in the Soviet Union where advanced diagnosis and treatments such as chronotherapy for insomnia and surgery for sleep apnea are performed. The journal *Sleep*, edited at Stanford, is widely read by Soviet sleep researchers.

*Sleeping Pills*

Inasmuch as scientists do not yet understand the biochemistry of sleep—what happens in the brain during the transition between wakefulness and sleep—it is not surprising that little is known about how sleeping pills actually work. At an insomnia national consensus workshop at the National Institutes of Health in Washington in 1982, one of the world's leading experts on psychopharmacology was asked, "How do sleeping pills work?" I was surprised to hear him answer, "I don't know." Furthermore, because there is no verified theory to explain the sleep process, few of the available sleeping pills have been specifically developed to promote sleep. Most hypnotics in use are derivatives of medications that were accidently discovered to have sedative effects—that is, they cause a general decrease in arousal.

The effectiveness and safety of many sleeping pills have been carefully studied in sleep laboratories. Several findings emerge: Most sleeping pills are initially effective in reducing sleep latency (time from lights out to the initiation of sleep) and in reducing nocturnal awakenings and increasing total sleep time. These same sedatives lose their effectiveness after two to four weeks, at which point the objective sleep parameters return to their baseline (prior to drug therapy) values. As the hypnotics lose their effectiveness, the patient's response is often to continually escalate the dose. Although many sleeping pills reduce slow-wave sleep and/or REM sleep, there is no specific associated ill health effect.

Abrupt withdrawal of sedative medications after chronic use causes a rebound insomnia. The brain becomes accustomed to a certain drug level, and rapid withdrawal may cause a night of total insomnia. Unfortunately, patients usually assume that this "rebound" is proof that they cannot sleep without medication, so they continue to use it. The trick is to withdraw very gradually and at the same time substitute sleep hygiene techniques for the missing drugs.

Sleeping pills are generally most effective for short-term, transient insomnia, as opposed to chronic insomnia. They are also useful as an adjunct to a sleep hygiene program. In chronic insomnia, non-drug techniques are more effective. This information is starting to reach physicians and the general population, and sleeping pill usage is on the decline. About two of every one hundred American adults will receive a sleeping pill prescription in the next year.

### Some Tangential Factors

*Sleep Environment*

**M**ost of us experience a poorer quality of sleep in an unfamiliar environment. For this reason, the data collected in a sleep research laboratory during the first night of study are often discarded. The inability to adapt immediately to the laboratory—the so-called first-night effect—typically causes lower quality sleep with more awakenings. After one night most subjects become acclimated to the laboratory environment, and their sleep patterns return to normal levels.

Insomniacs, on the other hand, typically overemphasize the importance of the sleep environment. Many fret and anticipate that their sleep problems will become more pronounced away from home—a fear that often causes them to forego vacations. Indeed, insomniacs will blame their sleep problem on bed partners, noises, lights, room temperatures, exercise, weather conditions, type of bed, nutrition, and anything else they can think of. Is there evidence for these assertions, or are they self-fulfilling prophecies? Let's see.

*Exercise*

If the purpose of sleep is to restore our worn-down body we would expect an increase in exercise to promote substantially deeper sleep. Such a straightforward correlation does not exist; individuals who exercise strenuously do not improve their sleep the following night. In fact, strenuous exercise in a nonathlete tends to increase physiological arousal, which results in disturbed sleep. This effect is even more pronounced if one exercises at an hour close to bedtime.

I rediscovered this truth recently on a ski trip to Colorado. I was not in very good shape but nevertheless skied a strenuous six hours on my first day at Aspen. That night my muscles were fatigued and aching. I was exhausted. Unable to fully relax my muscles and wind down, I had a poor night's sleep.

Regular exercise is a different story. Athletes who regularly exercise have more Stage 3-4 deep sleep than nonathletes. Furthermore, athletes' deep sleep is reduced when they stop exercising. For the average person, however, exercise or the lack of it probably does not play a big role in insomnia. To enhance sleep, exercise should be regular and performed in the late afternoon or early evening.

*Noise*

Insomniacs are correct about noise. It can disturb sleep. Light sleepers will probably wake up more readily than deep sleepers in response to the same noises. Light sleepers have higher body temperatures and faster heart and respiratory rates than deep sleepers. These characteristics may be innate or the result of chronic worrying and an inability to relax.

Whether a noise will actually disturb sleep and cause insomnia depends, too, on the type of noise, the duration of the noise, and when it occurs. Mothers can sleep through a thunderstorm but will awaken rapidly to their babies' crying. Sleepers wake up more rapidly if noises occur in Stage 1 or Stage 2 sleep than in Stage 3-4 sleep. Older people have less Stage 3-4 deep sleep than younger people and, not surprisingly, are more sensitive to noise. Noise in REM sleep also provokes rapid awakenings unless the noise is incorporated into a dream. Longer-lasting sounds, loud radios or televisions, are more likely to cause sleep disturbances than short but louder sounds, such as a firecracker going off.

Even though the sleeper may be unaware of a particular noise and sleep through it, it can still have an important effect on sleep quality and subsequent daytime functioning. When polysomnographs of the sleep of residents living near the Los Angeles airport were compared with those of other residents living in quieter Los Angeles neighborhoods, those in the airport area showed more widespread, lighter sleep and less slow-wave sleep, and their sleep disruption could be directly traced to aircraft fly-overs.

We can therefore conclude that there probably is a link between insomnia and noise. Some individuals may be genetically programmed to be lighter sleepers and more sensitive to noise. Chronic noise, however, will disturb the sleep even of a deep sleeper. Fiberglass earplugs can successfully block out noise such as street traffic sounds, which will result in increased slow-wave sleep. Studies of another strategy—the use of "white noise" (for example, air conditioning and fans) to block out disruptive sounds—have not proven it to be effective in enhancing sleep.

### Nutrition
"If I eat a nutritious, large, warm meal right before bedtime will I sleep better?" This question, which actually derives from Aristotle's ancient sleep theory that digestion and internal evaporation cause sleep, was addressed to me following a lecture I recently gave. The answer is no. Surprisingly, diet and nutrition have little effect on sleep. There is no specific food that will reliably induce sleep. On the other hand, excessive loss or gain of weight can change sleep quality. In general, loss of weight is associated with decreased sleep, and excessive weight with increased sleep. Some researchers claim that a balanced diet results in deeper sleep than a high-carbohydrate, low-fat diet. Others claim that the amino acid L-tryptophan, which naturally occurs in milk, has sleep-inducing properties. The L-tryptophan in milk may be sedating, or—because of its psychological associations—a glass of warm milk may be soothing. Or the type of milk may determine the effect. Horlick's Malted Milk was shown to promote sleep when taken near bedtime. For the insomniac, I usually advise avoiding spicy meals and foods with caffeine (colas, chocolate, tea, coffee) near bedtime, and beyond that not to worry about diet. Alexandre Dumas, the French author of *The Count of Monte Cristo*, suffered from intermittent insomnia. He was cured by his doctor's prescription that he eat an apple every morning at 7:00 A.M. under the Arc de Triomphe. Although it is doubtful the apple helped, the ensuing regular sleep-wake cycle in harmony with the light-dark cycle—and the belief he would be cured—actually worked.

### Beds and Bed Partners
An 1893 advertisement for twin beds advised readers that "Our English cousins are now sleeping in separate beds. The reason is: Never breathe the breath of another."

Nearly one hundred years later, most American couples continue to share the same bed. Some couples claim they cannot sleep without each other. In a laboratory study of couples who had been married for one to three years the claim was proven true. Sleep was lighter and more disturbed when husband and wife slept apart. Insomniacs, however, are often better off sleeping alone. Snoring, leg movements, and tossing and turning can disturb or even cause insomnia in a bed partner. The sleeping partner may resent waking up to console the anxious insomniac; conversely, the insomniac may bitterly resent the partner's peacefully sleeping even though the family may be sharing the same stressful lifestyle. In these situations, it is best to sleep not only in separate beds but in separate rooms.

The comfort of a particular bed is related more to cultural and personal preferences than to any intrinsic quality. Sleeping on a floor, a hammock, in space with zero gravity, on water beds, or in an upright position has little effect if one is accustomed to it. Sleeping on excessively hard surfaces, such as on a wooden or concrete floor (which I recently did while snowed in at the Seattle airport), does cause more awakenings and disturbed sleep. A variety of other sleep aides—custom beds, vibrating beds, or electric blankets—do not seem to enhance sleep. Since sleep is part of an internal biological rhythm, it is not surprising that outside environmental stimuli play a relatively minor role.

*Light and Dark*
The assertion of some insomniacs that their problems are related to light is supported by recent findings. Experiments with rodents and lower organisms have established that a few seconds' pulse of light can trigger an arousal response in the biological clock. In a number of animal species, including man, there is a direct pathway from the retina of the eye to a site close to the brain's biological clock in the hypothalmus. In environments of steady light or darkness, such as the Arctic region, sleep-wake cycles are generally disturbed in those unaccustomed to this pattern. Although research in this field is still in its infancy, we do know that insomniacs are best off making their rooms as dark as possible by blacking out windows, using clocks and radios without lights, and not turning on night lights. If they need to leave their beds at night they should use red light, a wave length that should not affect the biological clock. Upon awakening in the morning, they should pull up the shades, turn on lights, and try to obtain sunlight. One of my patients, a printer who worked a regular 4:00 P.M. to midnight shift, was cured of his early-morning-awakening insomnia by simply darkening his room.

*Weather, Temperature, and Altitude*
Some Californians claim that their sleep is worse when the dry Santa Ana winds prevail. Others claim that their worst sleep periods occur

when a particular weather front is approaching. There is some research evidence indicating that large swings in barometric pressures can influence the ability to fall asleep. However, I would not advise remaining awake at night to evaluate the weather forecast. The effects of climate on sleep are rather minimal.

Sleeping in a room in which the temperature is above 75 degrees Fahrenheit may disturb some sleepers; it can result in more awakenings and lighter sleep. At 100 degrees Fahrenheit sleep is unmistakably disturbed for almost everyone. On the other hand, the belief that a cool room (around 60 degrees Fahrenheit) improves sleep has not yet been validated.

Sleep at high altitudes is often disturbed because of the reduced oxygen level. When the brain senses reduced oxygen at night, increased arousals typically follow. After several nights of adaptation, however, sleep returns to normal value.

### Instructions to Improve Sleep in a Chronic Insomniac

"**E**arly to bed, early to rise, makes a man healthy, wealthy, and wise." This Ben Franklin adage suggests that being an early bird can contribute to success. In the days before electric lights this was probably true. At present most patients and doctors underemphasize the need for a regular sleep schedule whether it be an early-bird or late-rising schedule. The fact is, however, that the most important strategy for maximizing the circadian rhythm of sleep and wakefulness is to maintain a regular sleep-wake schedule, seven days a week. Chronic insomniacs should start with a restricted schedule, aiming for about 4 to 5 hours of sleep. Once sleep efficiency of about 90 percent is reached—sleeping 4½ hours out of the 5 allotted—the sleep schedule can be increased by 30 to 60 minutes.

Select a rigid, structured sleep schedule that approximates your maximum current total sleep time. If your sleep diary shows you only get 4 hours of actual sleep—not time in bed—then aim for a 4-hour schedule.

1 Into bed 1:00 A.M. Out of bed 5:00 A.M. (Set alarm).

**2.** Bedroom must be completely dark during sleep period; turn on lights and raise shades at rising time.

**3.** If awake and relaxed during sleep period stay in bed.

**4.** If awake and anxious during sleep period do relaxation techniques in bed. If still anxious, get out of bed.

**5.** If out of bed do household chores (laundry, cleaning, paying bills, etc.) If you feel sleepy get back into bed. Regardless of what time you return to bed, get up with the 5:00 A.M. alarm.

**6.** Avoid alcohol, caffeine, and cigarettes within 5 hours of scheduled bedtime. If possible, eliminate all of these substances. Chronic drug-dependent patients should withdraw one therapeutic dose of medication every four to five days.

**7.** Avoid exercise within 2 hours of bedtime. Exercise in the late afternoon may help sleep, but overall exercise is not a key factor in improving sleep. Neither are meals. Simply avoid spicy foods near bedtime.

**8.** If you have persistent active thoughts near or during bedtime, keep a cognitive diary. Set aside a 20-minute period after dinner when you try to worry as much as you can. Write the worries down, along with short-term or long-term solutions.

**9.** After two consecutive nights of poor sleep, based on your own diary, on the third night take one sleeping pill 30 minutes before your scheduled bedtime. In general, short-acting medications are appropriate, but longer-acting drugs may be more effective for patients with significant anxiety during the daytime. L-tryptophan can be used by those who prefer non-prescription aids.

**10.** Avoid naps.

Following these rules is difficult, but this course of treatment is successful in most patients within three to five weeks. Maintaining the sleep schedule is probably 75 percent of the battle. The sleep period can be lengthened gradually as success is achieved. Keeping a sleep diary for the entire three-to-five-week period is important and may in itself be curative. (When smokers are asked to keep a diary of their behavior, the number of cigarettes they smoke decreases.) Some side benefits accrue from following these rules: A friend whose wife is trying the program for her insomnia called to tell me "We have all the laundry done now!" A few weeks later her sleep was better, too. Those who are not helped by practicing all the rules for good sleep hygiene almost certainly have another diagnosis, such as psychiatric or physiologic insomnia.

## Key Questions in Diagnosing Insomnia

A few simple questions can help clarify the cause:

**1.** *How long have you had your sleep problem?* If less than three weeks, forget about it.

**2.** *Are you depressed?* If so, your sleep problem may just be a symptom of a mood disorder that needs psychiatric evaluation.

**3.** *Are you anxious about going to sleep?* If so, try the sleep hygiene techniques in this chapter.

**4.** *Do you snore loudly or pause in breathing at night?* If so, go to a sleep clinic for an evaluation.

**5.** *Do your legs kick or bother you throughout much of the night?* If so, go to a sleep clinic for an evaluation.

**6.** *Do you use alcohol or sleeping pills or other sedating drugs near bedtime?* If yes, these substances themselves could be causing your problem. Try the sleep hygiene techniques my patients have found useful, preferably under the supervision of your physician.

**7.** *Do you have a fairly regular, conventional sleep-wake schedule?* If not, this could be the cause of your insomnia. Try to adopt a regular schedule seven days a week.

Insomnia is not merely a nocturnal problem; it is a 24-hour problem. It was not until 1978, however, that a method was developed to evaluate the effects of sleeping pills on daytime functioning. Our nights influence our days; our sleep influences our performance. Sleeping-pill users should be concerned not only with how they are going to get to sleep but also how they are going to feel at work or play the next day. In the next chapter we will explore the other side of the coin—daytime alertness.

# 8

# From Deep Doze To Full Alert

At first glance the waiting room of a sleep clinic may easily be mistaken for the typical doctor's office. Half the patients are nervous, aimlessly picking up and quickly discarding magazines, impatiently waiting to see the doctor. The other half, however, are dead asleep, snoring away, sprawled out on chairs and sofas. These are the hypersomniacs.

Although disorders of excessive somnolence (hypersomnia) are not that common in the general population—identified in one to four people per hundred—these sleepy people represent 50 percent of all patients seen in over three hundred sleep clinics throughout the United States. One newspaper chain ran a story in 1982 entitled "A Nation of Sleepy Heads," based on this figure.

Who are all these sleepy people? Why are they rushing to sleep clinics when the insomniacs, who are much more numerous, are not? Maybe these hypersomniacs are sleepy because they have insomnia at night? With all these questions in mind, I go out to the waiting room, glance at my chart, and call for a Mr. Marcus. An attractive middle-aged woman elbows the stomach of a slightly overweight man, waking him up, and I am introduced to Mr. and Mrs. Marcus. Unlike some patients whom we awaken, Mr. Marcus stands up without any difficulty, shakes my hand, and acts alert, oblivious to the fact that he has just been snoozing away.

In the examination room Mrs. Marcus tells the story: "For I don't know how many years I've been telling him to go see a doctor. He falls asleep as soon as he starts reading the newspaper after work. Then he naps after dinner. He falls asleep watching TV. Sometimes he falls asleep when we're having company, but when I nudge him, he says he's been awake." Having been through similar scenarios, I'm curious to find out what brought the patient in at this time.

Mr. Marcus explains, "Well, I've always been a napper, but I could never understand why my wife said I should go to a doctor until last month. I've been a San Francisco Forty-Niner fan with season tickets for many years, during all their losing seasons. But last month I was at the championship playoff game, the most important game I've ever been to, and I missed it. I fell asleep in the second half. At that point I figured I'd lost control of my ability to stay awake and here I am."

Clearly, falling asleep at a special event of great personal importance seems extraordinary. Sleep clinicians label this "pathological sleepiness." But what about falling asleep while driving, at work, or during a boring lecture or concert? And on the other hand, what is optimal alertness? Feeling energetic all day or maybe for just a few hours after coffee break?

**D**istinguishing those patients who are abnormally sleepy from those who are not is a serious business for a sleep clinic. For example, in the 1960s and 1970s it was quite common for patients to feign abnormal sleepiness in order to obtain stimulant prescriptions, such as amphetamines. In California, patients who exhibit abnormal sleepiness must be reported to the Department of Motor Vehicles, which may then restrict or even recall their driving licenses. This emphasizes our responsibility for accuracy in diagnosis and treatment planning—both to protect the patient and society, and to protect ourselves from malpractice suits. Most importantly, abnormal sleepiness can be a debilitating illness. Imagine what your life would be like if, whenever you were awake, it felt like 3:00 A.M.!

In the early 1970s researchers at Stanford, led by Dr. William Dement, set out to develop a technique for accurately measuring alertness and sleepiness. The first attempt, a commonsense approach of asking patients how they felt, did not work very well. The very nature of drowsiness dulls the senses and decreases the possibility of making accurate assessments, even of one's own feelings. Mr. Marcus, for example, denied the presence of pathological sleepiness for many years. If you are accustomed to being drowsy you may not be able to discriminate between sleepiness and alertness.

The approach next indicated therefore was to develop a questionnaire, which was called the Stanford Sleepiness Scale. Patients had to rate their level of alertness on a scale of 1 to 7 every few hours. This scale was still an introspective-subjective measure, so that it had drawbacks similar to those of the direct questioning approach. Sleepy patients often continued to rate themselves "alert" when direct visual observation indicated otherwise.

Another approach was to monitor for 24 hours around the clock the EEG (brain waves) of those suspected of diminished alertness, noting how often they fell asleep. This technique was effective in experimental subjects but proved too laborious and expensive for an outpatient clinic. Dr. Dement and Dr. Mary Carskadon then devised a more practical approach. The more drowsy the patients, the more likely they should be to fall asleep rapidly at any time—given the proper conditions, such as a dark, quiet room and a comfortable bed. In 1978, Dr. Gary Richardson joined the research team and they published the first paper on the application of this concept to the testing of patients. The name coined for the new test was Multiple Sleep Latency Test, or MSLT.

This test is actually quite simple. Patients, or normal volunteers, come into the sleep laboratory during the daytime. A minimal number of electrodes are attached to the face and scalp in order to determine precisely the moment of transition from wakefulness to sleep. The onset of sleep is unambiguous when recorded electrophysiologically, so this determination is precise. The subjects are placed in a quiet, dark room at 10:00 A.M., asked to lie in bed and try to fall asleep. If they do not fall asleep within twenty minutes, the test is ended and a score of 20 assigned. If, on the other hand, they fall asleep within four minutes, the nap is terminated and a sleep latency score of 4 is assigned. The tests are repeated every two hours throughout the day and are usually halted after the 6:00 P.M. nap test. The MSLT has become the gold standard for accurate, scientific measurement of alertness.

### Narcolepsy

The first clinical application of the MSLT was with patients suffering from narcolepsy, an unusual sleep disorder characterized by irresistible, uncontrollable bouts of excessive sleepiness, automatic behavior (operating in automatic pilot mode, as illustrated in Chapter 1), and cataplexy (episodic muscle paralysis typically triggered by emotional situations). Narcoleptic patients sleep anywhere from 8 to 12 hours out of every 24 but find themselves falling asleep in some of the most bizarre circumstances. The most extraordinary case I encountered was a sixty-year-old trapper/hunter who left California at about the age of twenty and moved to Alaska. This intelligent patient had given up an academic career and moved north because of his inability to stay awake in classes and social situations.

As a trapper/hunter his primary targets were moose. Like many narcoleptic patients, he was often overtaken by a fit of cataplexy at the penultimate moment. Just when he had lined up a moose in his rifle sight and started to pull the trigger, he would collapse to the ground in a state of cataplectic sleep. Startled by the noise of the fall, the moose would run off. Eventually, however, the moose would return to investigate the source of the noise. The man would recover in time to see the moose standing over him; at that point he would awaken, grab his rifle, and shoot the moose. After this scenario had been repeated several times, the trapper claimed that he consciously used it as a technique for all his moose hunting.

This same trapper, who during the winter lived alone in an isolated cabin, confessed that he would become sleepy standing over the stove. He would drift into sleep and start to lose control, but remained conscious enough to see that he was going to fall and burn his body on the stove unless he put his hand out. His hand touching the stove woke him up.

Other patients have recounted bizarre sleepiness episodes. One patient was the expert handyman in a metropolitan hospital; everyone relied on him when anything needed to be repaired, from heart-lung transplant equipment to hospital beds. One day he brought his tools to a patient's room to fix a broken bed and apparently fell asleep. When he came to, he had welded each of his tools to the hospital bed. Another remarkable case was a farmer from New England. He came to our clinic after waking up in his tractor, amazed to find he had ploughed his entire field in a circle.

A host of other unusual symptoms are reported by narcoleptics: commuting home after work and driving fifty miles past their home exit before discovering their error; hallucinations of cars poised to attack them while they stop at a red light; collapsing to the ground just when getting ready to spank a child; falling asleep while eating—while actually lifting the fork to the mouth—or in the middle of a conversation. It's not surprising that people with these symptoms would believe themselves mentally ill or that physicians would diagnose them as having a psychiatric disorder. Fortunately, through the efforts of the American Narcolepsy Association and numerous sleep disorder clinics, there has been a tremendous increase in public awareness of narcolepsy. Today individuals with these clusters of symptoms are likely to be suspected of having narcolepsy.

Narcolepsy begins to make more sense when the Multiple Sleep Latency Test is applied. Given the opportunity to nap throughout the day, narcoleptic patients typically fall asleep in 2 to 3 minutes, or even less. Furthermore, they often go immediately from wakefulness straight into REM sleep. You will recall from Chapter 5 that it normally takes 90 minutes of sleep before the first REM period begins. Narcolepsy can be conceptualized as a failure to inhibit REM sleep. During the night the principal components of REM sleep—dreaming (hallucination) and inhibition of reflexes (paralysis and loss of muscle control)—are normal events. In narcolepsy, REM sleep intrudes into wakefulness, causing the startled patient suddenly to experience those states that are better left to a nocturnal REM period.

There is currently no effective treatment for narcolepsy. Improving or increasing nocturnal sleep does not help daytime alertness in narcoleptics. Stimulant medications are typically prescribed; these may, or may not, be helpful, depending on the severity of the symptoms. Cataplexy—loss of muscle control associated with expressions of emotion, especially anger and laughter—can be controlled to some extent with medication. Many narcoleptic patients, however, learn to control their emotions in order to avoid cataplectic episodes. Thus, flat affect and avoidance of emotionally charged situations are not uncommon among these patients.

Narcolepsy is a genetically transmitted disease, probably caused by an imbalance of one or more neurotransmitters (brain chemicals) in the part of the brain that controls REM sleep. Abnormalities of the body's immune system may also be involved. The onset of the disease usually occurs during puberty and is essentially lifelong, with little change in symptoms. Relatively rare, it occurs in fewer than 1 in 1000 individuals. Its heritability is less than that of schizophrenia, asthma, or cleft palate but equivalent to that of coronary artery diseases. Since it is relatively rare, even children of a narcoleptic patient have only about a 5 percent chance of developing either narcolepsy or excessive somnolence.

Animals, too, suffer from narcolepsy. The Stanford Sleep Research Program, in trying to understand the disease and to devise more effective treatment, has worked with a colony of narcoleptic dogs. With dogs, the diagnosis is arrived at not by using the MSLT but by a performance test. Hungry dogs are led to a starting line, where pieces of meat are stretched out every few feet for the length of a hallway. The animals are timed to see how long it takes them to eat all the meat. The normal dogs are able to eat all the pieces within a matter of seconds. The narcoleptic dogs fall asleep and are overtaken by cataplexy en route and may take up to half an hour to complete the task.

## Sleep Apnea

The other major cause of excessive daytime sleepiness is *sleep apnea* (apnea is from the Latin, meaning "without breath"). I repeat this definition because I recall observing a sleep clinician guiding representatives from the United Nations on a tour of his laboratory. After a 30-minute demonstration of the equipment used for monitoring this potentially life-threatening illness, one of the U.N. representatives finally raised his hand and asked, "What is apnea?"

Charles Dickens's description of the "Young Dropsy," a fat, somnolent, red-faced boy in *The Posthumous Papers of the Pickwick Club* is one of the earliest accurate descriptions of a variant of sleep apnea. At one point this syndrome actually carried the misnomer "Pickwickian Syndrome."

> A most violent and startling knocking was heard at the door; it was not an ordinary double knock, but a constant and uninterrupted succession of the loudest single raps, as if the knocker were endowed with perpetual motion, or the person outside had forgotten to leave off.
>
> Mr. Lowten hurried to the door. . . . The object that presented itself to the eyes of the astonished clerk was a boy—a wonderfully fat boy . . . standing upright on the mat, with his eyes closed as if in sleep. He had never seen such a fat boy, in or out of a traveling caravan; and this, coupled with the utter calmness and repose of his appearance, so very different from what was reasonably to have

been expected of the inflicter of such knocks, smote him with wonder.

"What's the matter?" inquired the clerk. The extraordinary boy replied not a word; but he nodded once, and seemed, to the clerk's imagination, to snore feebly. "Where do you come from?" inquired the clerk.

The boy made no sign. He breathed heavily, but in all other respects was motionless.

The clerk repeated the question thrice, and receiving no answer, prepared to shut the door, when the boy suddenly opened his eyes, winked several times, sneezed once, and raised his hand as if to repeat the knocking. Finding the door open, he stared about him with astonishment, and at length fixed his eyes on Mr. Lowten's face.

"What the devil do you knock in that way for?" inquired the clerk, angrily. "Which way?" said the boy, in a slow, sleepy voice.

"Why, like forty hackney-coachmen," replied the clerk.

"Because master said I wasn't to leave off knocking till they opened the door, for fear I should go to sleep," said the boy.[21]

Sleep apnea refers to a variety of syndromes and is characterized by cessation or significant diminution of breathing during sleep. In the classical case the patient falls asleep quickly at night—as soon as the lights are turned out. Then, as sleep commences, there is a pause in breathing that lasts 30 seconds, or even as long as a minute. Frequently the respiratory muscles of the diaphram and abdomen make an increasingly strenuous effort to push air through the trachea. But the trachea is partially closed off, so that the effort is comparable to a musician's blowing into a partially obstructed trumpet. The resulting sounds of grunting, snorting, gasping, and onerous snoring increase. The heart may slow down or, in severe cases, skip a few beats. Receptors in the brain, sensing a loss of oxygen, struggle to awaken the sleeper. The unaware patient wakes up for 5 to 10 seconds, takes a deep breath, rolls over, and goes back to sleep, only to start another cycle of apnea.

These cyclical apneas frequently repeat four to five hundred times a night. More often than not, sleep apnea victims have no idea that

"The Fat Boy"

they are not breathing or that they are continually waking up. The brief awakenings, or arousals, are not remembered the next morning. Daytime sleepiness, however, is a common complaint. The sufferer may also mention that a bed partner complains of loud snoring noises. In fact, in the typical case a wife brings her middle-aged husband in for treatment, as Mrs. Marcus did.

One apnea patient reported that when he was in the military and sleeping in the barracks, he would frequently wake up with a sea of army helmets strewn around his bed. At different times throughout the night his barracks mates had flung their helmets at him to get him to stop snoring. (Sometimes a good shot would make him change positions, which would temporarily reduce his snoring.) Of course, the patient slept through it all. Later on in life, when he was on a business trip, he often had to rent the hotel rooms adjacent to his own to avoid complaints about his loud snoring.

Sleep apnea is now recognized as a cause of morbidity and sudden death. It is interesting to speculate why this disorder was not widely recognized until as recently as 1965. Perhaps because sleep has historically been considered an off-state of the human organism, in which all physiological functions operate at a reduced but steady level, few researchers thought the study of breathing during sleep worthwhile. No one was willing to stay awake at night to observe breathing patterns. (Actually, with a sleep-apneic patient, you simply need to put a mirror next to the mouth and nose to see if vapor condenses, indicating the presence or absence of air movement.) Another stumbling block was the common wisdom preserved by physicians as the answer to all sorts of medical problems: "Get a good night's sleep." This adage reflected the Aristotelian concept that sleep is a restorative process caused by the build-up and excretion of a hypotoxin. Given these assumptions, it is not surprising that the actual dangers of sleep have been ignored.

Although severe sleep apnea may occur in only 1 percent of the general population, milder forms of apnea are much more common in adult males and in the elderly. Several field studies of healthy senior citizen populations recently conducted by means of ambulatory home recordings (devices to monitor sleep and breathing were brought to the volunteers' homes) have revealed that over a third of the seniors had sleep apnea. That is, in people over age 65 it is quite common to have at least five breathing pauses during every hour of sleep.

If sleep apnea is so common in the elderly perhaps it is not a disorder. Perhaps it is nature's way of allowing us to "die peacefully during sleep." Several studies have reported that hospital deaths most frequently occur at night. Although these figures may be influenced by a lower level of nighttime alertness on the part of the nursing and house staffs, some sleep clinicians have advocated not treating seniors who exhibit sleep apnea. Current research indicates that hypoxia (lack of oxygen) caused by sleep apnea creates increased confusion upon awakening in the morning; this data can be extended to the assumption that chronic oxygen deprivation might exacerbate dementia and/or Alzheimer's disease.

Another threat to the elderly is posed by sleeping pills. Their use is extremely widespread among older people; about 40 percent of all hypnotic prescriptions written are for people over 60, although this group comprises only 15 percent of the general population. Furthermore, several studies indicate that 50 to 60 percent of hospital inpatients receive a hypnotic during their hospital stay. The catch is that sleeping pills change the arousal threshold, preventing the brain from awakening the sleeper during an apnea. As a result apneas become longer and more frequent. The combination of sleep apnea and sleeping pills may therefore be dangerous. The American Cancer Society's epidemiological study found that elderly individuals who use sleeping pills have a significantly greater mortality rate than those in the same age group who do not use hypnotics.

## Treatment of Sleep Apnea

Historians of ancient Greece described treatments for sleep apnea: "Dionysius, the son of Cleandrus . . . gradually became overloaded with flesh by reason of the luxury and gluttony in which he lived daily; hence, because of his obesity, he was afflicted with shortness of breath and fits of choking. So the physician prevailed that he should get some pine needles, exceedingly long, which they thrust through his ribs and belly whenever he happened to fall into a very deep sleep."[22]

Although obese men are at highest risk for sleep apnea, a number of other risk factors can be causative in non-obese patients: enlarged tonsils or tongue, excessive fat tissue in the throat, or a combination of other anatomical abnormalities that create a narrow airway. It is therefore not surprising that treatments range from surgery to medication, tongue-restraining devices, weight loss, changing sleep position (sleeping off the back), or wearing a small nasal pressure mask to keep the airway open. Increasingly sophisticated diagnostic techniques, such as filming the location of occlusion in the sleeping patient, allow sleep clinicians to match the correct treatment to the specific type of apnea. All of these newer treatments must be compared with the standard effective treatment, tracheostomy—placing a small hole and tube in the trachea, which the patient opens at night, in order to bypass the obstruction in the throat.

# Snoring

## Embarrassment and/or Risk Factor

Snoring can be a trivial annoyance or a symptom of a life threatening disorder. In the obstructive sleep-apnea syndrome, loud repetitive snoring is one of the cardinal diagnostic signs. Snoring is due to a narrowing of the airway passage, or stenosis, that occurs only during sleep, which is usually a result of reduced muscle tone or an anatomical peculiarity. The vibrations of air against the soft palate cause the actual sounds. If snoring is light and intermittent there is no abnormal effect on the lungs or respiratory system. If snoring is loud and continuous, however, even without apnea, respiration becomes abnormal and high blood pressure can develop during the night.

The amount of snoring varies from night to night. Sleep deprivation, alcohol, high altitudes, and certain sleep positions can all increase it. Whereas the full-blown sleep-apnea syndrome involves an almost complete occlusion of the airway, snoring indicates a moderate narrowing of the airway.

Snoring can occur in any age group, from infants to the elderly, and can vary from the benign to the serious. The most common and innocuous is nasal snoring, a low pitched nasal sound, similar to a nasal speaking voice. This type is often intermittent and is associated with the occurrence of the common cold or allergies. Many people don't realize they snore, but waking up with a dry mouth in the morning is often a telltale sign. Another pattern, oral-pharynx snoring, is more serious and is often associated with snorting and gurgling sounds. The more frequent these sounds, the greater the likelihood that sleep apnea may develop.

The epidemiological survey in San Marino found that 19 percent of the general population were habitual snorers (persons who reported snoring every night). Sex and age were important determinants. After age forty, 30 percent of males and 20 percent of females snored; after age sixty, the figures rose to 50 percent and 40 percent respectively. Weight was an important factor; about one-third of overweight and obese individuals were snorers, while only 10 percent of persons of normal weight were snorers. High blood pressure was more prevalent in snorers than non-snorers. After age, snoring and body weight appear to be equally good predicters of the presence of hypertension.

To sum up, snoring can be taken as a sign of mildly impaired respiration, which, if habitual, can lead to hypertension or sleep apnea. Snoring is also a social problem, embarrassing to the snorer and often disruptive to bed partners. The best cures for habitual snoring are weight loss or, if that fails, an operation called uvula-palatopharyngoplasty (a long name for a fairly routine procedure), in which the airway is widened by removing excessive fat tissue and part of the soft palate. Even a tiny amount of weight gain that results in more fat tissue in the throat can cause snoring.

### Normal Sleepiness

**M**ost of us will not suffer from narcolepsy or sleep apnea, the two most common causes of daytime sleepiness. Yet complaints of fatigue, lack of alertness, and sluggishness are common. How can we maintain maximum alertness? The MSLT has proved useful in answering this question. A large number of normal volunteers, those without sleep disorders, have

FIGURE 8-1

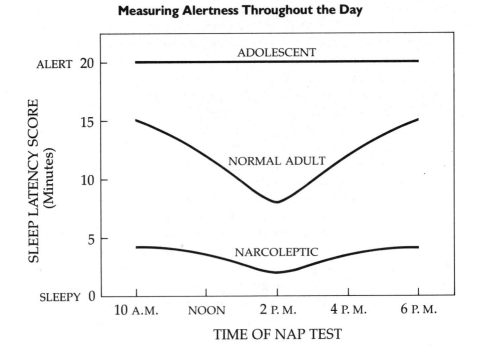

Measuring Alertness Throughout the Day

donated their daytime hours to help Stanford researchers measure alertness. The average adult, when given a chance to nap every 2 hours across the day, will fall asleep at several opportunities. They do not fall asleep within 2 to 3 minutes as narcoleptics do, but may average a 12-minute sleep latency score for the day. Sleep latency is the amount of time elapsed between lights off and actually falling asleep (see Figure 8-1).

Age makes a difference in alertness. Pre-adolescent children (ages ten to twelve) are the most alert humans. No matter at what daytime hour they are given an opportunity to nap, they don't fall asleep. Anyone who has observed this age group at a school playground during a break will confirm these sleep lab results. For contrast, observe the activities of adults during a break. Middle-aged and elderly people tend to be sleepier than young adults; given the right environment, there's a good chance they will nod off.

Several factors may account for these findings: Pre-adolescent children have one long, regular sleep period every day. Because disruptive social cues are absent (bed partner, late-night television, getting up early for work) pre-adolescents probably get the optimal amount of sleep each night—about 9 to 10 hours—and maintain a regular circadian rhythm, keeping the same sleep schedule seven days a week. In addition, this age group has a high percentage of Stage 3-4 sleep, which appears to have restorative value in maintaining subsequent alertness.

By contrast, the typical teenager and young adult have an irregular sleep-wake schedule, restricted total sleep time, and a gradual decline in slow-wave sleep. These factors become even more pronounced in middle age, when work schedules, child care, late-night TV, and social events may lead to mild, chronic sleep deprivation. Indeed, several sleep researchers believe that most Americans suffer from mild but chronic sleep deprivation.

If you place a group of college students on a regular 8-hour nighttime sleep schedule, their MSLT scores the following day will average about 15, meaning they fall asleep during some nap tests and not others. Remember, a score of 20 indicates maximum alertness and a score of zero indicates maximum sleepiness. But if nighttime sleep is increased to 10 hours for several nights, MSLT scores rise to 20—no falling asleep during nap tests. Conversely, if nighttime sleep is restricted to 4 hours a night for two nights, daytime MSLT scores drop to around 5, a sleepiness score close to the narcoleptic range.

Complete nocturnal sleep deprivation—no sleep at all—results in maximum sleepiness the next day, with volunteers falling asleep within 1 minute at each nap opportunity—the equivalent of pathological sleepiness. Following complete sleep deprivation it seems to take two nights of 8-hour "recovery" sleep before daytime alertness scores return to their baseline levels. As we saw in the studies reviewed in Chapter 5, difficulties in performing daytime tasks occur only after complete sleep deprivation and not after the 4-hour sleep restriction.

These MSLT studies suggest a simple relationship between nocturnal sleep and daytime alertness. The more sleep at night, the greater the daytime alertness; the less sleep at night, the less the daytime alertness. Our results indicate that due to reduced nocturnal sleep, disrupted sleep schedules, and perhaps the normal aging process (reduced Stage 3-4 sleep), most adult Americans are functioning below maximum alertness. When motivation is insufficient to overcome drowsiness, this may translate into poor performance.

To understand alertness, a further complicating factor must be taken into account. On almost all MSLT studies—of narcoleptics, apneics, normal volunteers, etc.—the shortest sleep latencies occur in the afternoon, typically at 2:00 P.M. At this time almost all subjects except pre-adolescents can easily fall asleep, often within several minutes. The so-called "post-lunch dip" is probably not due to the noon meal; it does not occur after breakfast or after dinner. This decline is a real physiological phenomenon and has been validated by many research groups. Most likely an ultradian rhythm—biological rhythm of less than 24 hours—of performance or alertness, independent of total nighttime sleep, is responsible for this afternoon dip. Not only are there circadian biological rhythms, those that fluctuate on a daily basis, but the presence of a 90-minute cycle of rest-activity has been hypothesized.

The impact of biological rhythms upon daytime alertness has been convincingly demonstrated by using the MSLT. When MSLTs are given across the 24-hour day, every 2 or 3 hours, subjects will show maximum sleepiness around 3:00 A.M., with a secondary dip around 2:00 P.M. There are several major determinants of daytime alertness:

(1) Circadian phase, or the match between internal time and external time: A day worker will be sleepy during a 3:00 A.M. sleep latency test, whereas a night worker should be alert at that time
(2) Total sleep time the night prior to the test

(3) The amount of Stage 3-4 sleep

(4) Regularity of sleep and work schedule

(5) Presence or absence of a sleep disorder such as narcolepsy or sleep apnea

Now that we have established that the average adult is functioning below the alertness level of a ten-year-old, our analysis of adult behavior should make more sense. Falling asleep at a boring 2:00 P.M. meeting would be normal. If you are the speaker at the meeting, falling asleep would indicate pathological sleepiness, suggesting the presence of sleep apnea or narcolepsy. Motivation should be sufficient to overcome sleepiness in that situation. Similarly, falling asleep on a long-distance drive at 3:00 A.M. would not be unusual, especially as it is more difficult to remain motivated during routine vigilance tasks.

Driving a motor vehicle is one of the best performance tests, as it requires sustained attention, although it is a fairly routine task. Listening to a talk show or music on a car radio reduces the tendency to fall asleep and makes for better driving. Driving in hot environments (90° F) leads to increased errors such as lane drifting. But it is not the car itself, the heat, or the lack of music that causes drowsiness. The underlying physiological sleepiness is simply coming forth. A physiologically alert person will not fall asleep even in a hot car without a radio.

In certain cultures, recognition of these naturally occurring physiological drops in alertness are built into society's schedule—hence the siesta. In the United States and other industrialized nations, caffeine has become a replacement for the siesta. Drinking a cup of coffee in the morning and in the afternoon is a common method for combating drowsiness. Surprisingly, there have been few MSLT studies measuring the effects of caffeine on alertness. Preliminary results suggest that caffeine—three cups at 9:30 A.M.—promotes daytime alertness, especially if you are slightly sleep-deprived. If you are already sleeping 9 to 10 hours at night, caffeine will have little effect in promoting daytime alertness; you are probably already sufficiently alert. In practice, when caffeinated beverages are consumed on a regular basis, the alerting effect is diminished.

Caffeine is the most widely used stimulant in the world. It is found in coffee, tea, colas, chocolate, and in certain over-the-counter drugs, such as aspirin combinations, dieting aids, and alertness aids (e.g., No-Dōz). The average cup of coffee contains between 100 and 150 milligrams of caffeine; a 12-ounce cola contains between 35 and 55 milligrams; and a chocolate bar may be as high as 25 milligrams per ounce.

Caffeine exerts its main physiological effects on the central nervous system, the heart, and to a lesser extent the kidneys. With as little as 100–200 milligrams, the equivalent of one or two cups of coffee, caffeine's effect upon the brain, experienced as increased stimulation, is apparent. This may result in increased mental alertness without the disruption of coordination or thinking that occurs with more potent stimulants like amphetamines or cocaine. In everyday usage, however, caffeine may not provide the stimulating effect scientists have measured with single doses.

Like other stimulants, the increased arousal and performance is followed by a period of behavioral depression. Caffeine can have a paradoxical effect, especially when used in large quantities on a daily basis (such as five to ten cups per day). As the brain becomes accustomed to a certain level, more and more is needed to prevent the depression. This mental depression is associated with decreased performance and sleepiness.

The most effective way to obtain an alerting effect from caffeine is to use it only occasionally. It is not easy to stop a habit of regular coffee use. The use of coffee to get going in the morning reflects the effects of poor sleep-wake schedules and inadequate sleep rather than any defect in our biological system. The biological system is geared to help us wake up naturally if we maintain an adequate sleep-wake schedule.

### Sleeping Pills, Antihistamines, and Alertness

Adding to your sleep time does not necessarily make you more alert if your sleep was drug-induced. The most commonly prescribed sleeping pill in the 1970s was *flurazepam*. It became popular because it was safer than barbiturates and did not significantly reduce REM or Stage 3-4 sleep. It was not until the development of the MSLT that sleep researchers had the opportunity to accurately measure alertness following a good night's sleep with a hypnotic. Flurazepam is active in the central nervous system for at least 80 hours, so that it actually promotes daytime sedation, an effect rarely desired except for extremely anxious patients. That is, its hypnotic effect is equally potent at midnight and at 10:00 A. M. the next day. MSLT studies show that following a night of flurazepam treatment that has resulted in a good nocturnal sleep, patients are quite commonly sleepier than they would have been following their usual night of insomnia. Some of the shorter-acting hypnotics appear to be just as effective in inducing sleep without causing any daytime sleepiness. The objective measurement of drug effects is an important clinical tool. Patients' reports are not a reliable source of data. Patients will often deny, for example, that they are sleepy; happy to have gotten through the night without awakening, they don't want to be taken off their medication.

Antihistamines are one of the most commonly prescribed over-the-counter medications and are often self-prescribed sleeping aids. The standard antihistamine, *diphenhydnamine*, in use since the 1940s, is effective in relieving hives, allergies, nasal stuffiness, etc., but it is also known to be sedating. In fact, this compound is used in many over-the-counter sleeping pills. In the past few years new antihistamines, with a different chemical structure, have been marketed in Europe, Canada, and, most recently, the United States. They are supposed to be devoid of sedative effects and thus beneficial to daytime users. At Stanford we have been testing the sedative effects of some of the new and older antihistamines. So far we have found that the older antihistamines produce a level of sedation that is equivalent to getting about 4 hours of sleep at night. At certain doses, the newer drugs did not produce any sedation. Regardless of whether volunteers received old or new antihistamines or none at all (placebo), everyone was equally sleepy during the 2:00 P.M. MSLT naps, the afternoon dip.

## Looking Ahead

I f you think back to the train crash or the airline incidents described earlier in this book, their causes are now, I hope, clear to you. Despite the greatest will power, our biology is geared to make us sleep on a daily basis; its force is difficult to fight. Since Edison's invention of the light bulb and the start of the jet age, the distinction between night and day has become blurred. How will we deal with the conflict between our physiology and our increasing need to safely and efficiently operate around the clock?

We are confronted with a serious problem. In a survey of over 900 rotating shiftworkers, I found that 56 percent admitted to falling asleep or nodding off at work on a regular basis. These sleepy workers were not lying down in a secret rest area but were falling asleep while performing their jobs in paper mills, chemical plants, refineries, and manufacturing plants. To convince industry of the dangers inherent in shiftwork has become easier in recent years. In talking to a group of shiftworkers at a large plant in upstate New York and explaining the biological clock, I shared with them the day's lead story in their own local newspaper, which I happened upon by chance: "Nuclear Plant Firemen Caught Napping While on Duty." At 3:00 A.M., the paper reported, supervisors made an unannounced inspection and found five of the thirteen employees sleeping. The plant claimed that this did not constitute a serious safety hazard, and the U.S. Nuclear Regulatory Commission was not notified and had no record of the event. The people who worked next door to the nuclear plant did not quite agree with the plant management's assessment. Looking back at Three-Mile Island, Bhopal, and Chernobyl, it is becoming clear that this is a critical issue.

Looking ahead, I see three basic strategies. At one extreme, we can ignore our biological clocks and expect humans to perform like superhumans—to maintain peak alertness, and to live with the risks. At the other extreme, we can try to turn our society back to a diurnal one. In Japan, for example, women are not allowed to perform night work, and the Japanese propose to operate industries around the clock only if a continuous process is essential. A nuclear power plant would qualify, but not an automobile plant, whose goal would be merely to improve capital investment. Both these choices—ignoring biological clocks or trying to go back to pre-Edison times—are unrealistic.

A better approach is to try to acknowledge and integrate information on biological rhythm into our round-the-clock society. In some industries this is already happening. One of the positive outcomes of the Burlington Northern train crash was the realization that a number of sophisticated alerting devices are available and are being used on some railway lines. These devices will automatically stop the train if the engineer falls asleep or fails to respond to restrictive signals.

In the skies, other approaches will be needed. The airlines have generally been less amenable to including biological rhythm as a factor in their operation scheduling. Pilots are pretty much supposed to be able to handle any situation. As continuing deregulation brings an increase in public concern with airline safety this will change. Recently I received an inquiry from a captain who flies for a major airline. He wanted to know how his company could improve scheduling and safety techniques for their pilots. Maybe because this airline is employee owned, they are more willing to face the fact that pilots become fatigued, and that pilot fatigue is dangerous for crew and passengers.

Another way out of this dilemma is to develop a simpler method of resetting our biological clocks—something as simple as resetting your watch after crossing a time zone. If switching our watches ahead by one hour could also switch all our biological rhythms we could be wide awake whenever we chose to be. This option may come about through development of a wonder drug, or perhaps millions of years of evolution in our new time environment may alter our biological clocks. In the meanwhile we have to adapt our old-fashioned biological system to a world that never stops.

As a sleep researcher, I take some satisfaction in knowing that by applying what we do know we can improve safety and health. The manager of a utility plant in northern California called me to discuss shiftwork schedules. As usual, I asked what schedule his plant was on. When he replied, "The Stanford Schedule," my anxiety level soared. So many similar versions of the poorly designed traditional schedules are known by so many different names, I wondered if he had attached the Stanford name to one of those. Striving to remain calm, I asked, "Which schedule is that?" His answer was, "The one that rotates in the right direction"—one of the schedules we had introduced at the GSL plant in Utah! But they had not adopted the *rate* of rotation of the better of the two GSL schedules! Change comes slowly, but so long as it moves in the right direction, we can look forward to putting the findings of chronobiology at the service of improved health and safety.

# Notes

1. William C. Dement, *Some Must Watch While Some Must Sleep* (Stanford, CA: The Portable Stanford, Stanford Alumni Association, 1972), pp. 8–12.
2. As quoted in Stefan Lorant, *Pittsburgh: The Story of an American City* (Lenox, MA: Authors Edition, Inc., 1964), p. 214.
3. *Ibid.,* p. 281.
4. The Interchurch World Movement, Commission of Inquiry, *Report on the Steel Strike of 1919* (New York: Harcourt, Brace and Howe, 1920).
5. S. Adele Shaw, "Now That Jerry Has Time To Live," *Survey* (Sept. 1, 1924): 568.
6. House Committee on Investigation of United States Steel Corporation, *United States Steel Corporation Hearings,* vol.14, 62nd Cong., 2nd sess., 1912, p. 2841.
7. House Committee on Science and Technology, Hearings Before the Subcommittee on Investigations and Oversight, *Biological Clocks and Shift Work Scheduling,* 98th Cong., 1st sess., 1983, pp. 4–5.
8. *Ibid.,* p. 170.
9. *Ibid.,* pp. 12–13.
10. Wiley Post and Harold Gatty, *Around the World in Eight Days: The Flight of the Winnie Mae* (New York: Rand McNally, 1931), p. 27.
11. *Ibid.,* p. 119.
12. Doug Beal, *Spike: The Story of the Victorious U.S. Volleyball Team* (San Diego, CA: Avant Books, 1985), p. 133.
13. Laverne C. Johnson, in *Sleep: Physiology and Pathology, A Symposium,* ed. Anthony Kales (Philadelphia: J.B. Lippincott Company, 1969), p. 206.
14. Summary position paper, Soviet Sleep Symposium, 1982.
15. As quoted in Daniel Goleman, "Staying Up," *Psychology Today* (March 1982): 27. Reprinted with permission of *Psychology Today.* Copyright 1982 (APA).
16. C.G. Jung, *Dreams* trans. R.F.C. Hull (Princeton, NJ: Princeton University Press, 1974), p. 27.
17. Sigmund Freud, *The Interpretation of Dreams* trans. A.A. Brill (New York: Random House, 1950), pp. 247–248.
18. From Hervey de Saint-Denys, *Les Rêves et les Moyens de les Diriger,* as quoted in Charles McCreery, *Psychical Phenomena and the Physical World* (London: Hamish Hamilton, 1973), pp. 102–104.
19. W.C. Dement and E.A. Wolpert, "Relationships in the Manifest Content of Dreams Occurring on the Same Night," *Journal of Nervous and Mental Disease,* as quoted in David Foulkes, *The Psychology of Sleep* (New York: Charles Scribner's Sons, 1966), pp. 95–96.
20. Christian Guilleminault, Peter Pool, Jorge Motta, and Anne M. Gillis, "Sinus Arrest During REM Sleep in Young Adults," *The New England Journal of Medicine* vol. 311, no. 16 (Oct. 18, 1985): 1006–1007.
21. Charles Dickens, *The Posthumous Papers of the Pickwick Club* (London: Oxford University Press, 1948), p. 753.
22. Athenaeus, *The Deipnosophists,* trans. Charles Gulick, vol. 5 (Cambridge, Mass.: Harvard University Press, 1863), pp. 491, 497.

# Glossary

**asystole:**  Brief episode of complete cessation of heartbeat.

**automatic behavior:**  Behavior enacted in a diminished state of alertness. The person is typically unaware of the actions carried out during this state.

**biological clock:**  An innate physiological system capable of measuring the passage of time in a living organism.

**biological rhythm:**  Oscillations in physiological systems that are caused by a biological clock. Such oscillations need not be driven by outside environmental stimuli.

**biorhythms:**  Fluctuations in human behavior attributed to the timing of one's birth. There is currently no evidence that biorhythms exist.

**brain stem:**  One of the three main subdivisions of the brain. Located at the base of the brain, it connects the spinal cord with the rest of the brain, and is thought to contain the mechanisms that regulate sleep-wake behavior.

**cataplexy:**  A sudden attack of complete or partial muscular paralysis precipitated by strong emotion. One of the cardinal symptoms of narcolepsy.

**chronobiology:**  The branch of science that studies the timing of biological systems.

**chronotherapy:**  Technique for enabling a patient to adjust to an earlier bedtime by progressively delaying bedtimes to later hours.

**circadian phase:**  About a day. Circadian rhythms are biological rhythms that oscillate on or about a daily (24-hour) basis.

**circannual:**  Biological rhythms that oscillate on or about a yearly basis, such as hibernation in animals.

**deep sleep or slow-wave sleep:** Stages 3 and 4 of NREM sleep, also called delta sleep because of the appearance of high amplitude, low frequency (slow) brain waves known as delta waves.

**diurnal:** An organism that is active by day, asleep by night.

**dream telepathy:** The concept that awake individuals can influence the dream content of sleeping subjects through psychic powers.

**EEG (electroencephalogram):** The recording of brain activity made by an electroencephalograph. Tiny electrical currents in the brain are recorded with small scalp electrodes and amplified onto a polygraph.

**electrosleep:** Technique heralded in the 1960s in the Soviet Union which attempted to improve sleep by sending tiny electrical impulses to the brain in a pattern that mimics sleep.

**enuresis:** Bed-wetting, a sleep disorder occurring mainly in young children.

**flurazepam:** Generic name for the commonly prescribed sleeping pill Dalmane. Flurazepam has a very long action and can produce daytime sleepiness.

**free-running:** Phenomenon appearing when an organism is not governed by time cues; it will adjust to its own unique day length.

**Halcion:** Short-acting sleeping pill that is useful in transient insomnia and in adjusting to jet lag.

**hypersomnia:** The complaint of too much sleep or of feeling sleepy all the time, despite a normal amount of sleep at night.

**hypotoxin:** A hypothetical chemical substance that builds up in the body and then is depleted, supposedly responsible for sleep and wakefulness. Despite the widespread belief in this model, there is little data to support it.

**hypoxia:** Below-normal levels of oxygen in the blood.

**infradian:** Biological rhythms that oscillate on a cycle much longer than 24 hours. A menstrual cycle is an infradian rhythm.

**insomnia:** Complaint of too little sleep, difficulty falling asleep and staying asleep, frequent awakenings, and/or poor quality sleep.

**internal desynchronization:** State induced by lack of, or erratic, time cues, during which physiological rhythms that are usually linked go out of phase.

**jet lag:** Brief maladjustment experienced when a change of time zones causes biological rhythm to become out of phase with new local time.

**lucid dream:** Experience of being aware during a dream that one is dreaming.

**MSLT (Multiple Sleep Latency Test):** A practical physiological test developed to measure alertness and sleepiness.

**microsleep:**  A brief intrusion of sleep in an awake individual, usually lasting only a few seconds and often associated with a lapse in normal waking behavior.

**NREM (non-REM) sleep:**  All sleep stages except REM sleep; includes stages 1, 2, 3, and 4. In NREM sleep there are No Rapid Eye Movements.

**narcolepsy:**  An illness characterized by excessive sleepiness and attacks of muscular weakness (cataplexy). Sleep paralysis, hallucinations, and disrupted nocturnal sleep are frequently associated symptoms.

**neuron:**  A specialized cell that forms the basic unit of the nervous system.

**neurotransmitters:**  Chemicals in the brain that allow one neuron to com municate with another neuron.

**night terrors:**  Awakenings occurring during non-REM sleep, usually stages 3 and 4, characterized by loud screams, feelings of impending doom, perspiration, racing heartbeat, and glazed eyes. May be precipitated by life stress.

**night-shift paralysis:**  A temporary inability to move, experienced by night-shift employees in sedentary positions.

**nightmare:**  A vivid, anxiety-provoking dream often occurring near time of awakening and usually occurring during REM sleep.

**oscillate:**  To swing or move to and fro as a pendulum does; to fluctuate between states—for example, from high to low body temperature.

**"over the top":**  Term describing air flights over the Arctic Circle which result in the crossing of many time zones.

**phase advance:**  Shifting of biological rhythms to earlier hours.

**phase delay:**  Shifting of biological rhythms to later hours.

**polygraph:**  An instrument able to simultaneously record tracings of several different physiological variables.

**polysomnography:**  Continuous and simultaneous recording of physiological variables during sleep.

**pseudo-insomnia:**  Complaint of disturbed or too little sleep despite findings of normal sleep patterns in objective all-night sleep recordings.

**REM sleep (Rapid Eye Movement sleep):**  Active sleep state during which brain activity approximates the activity found in the alert waking state, spinal reflexes are inhibited (paralysis), and vivid dreaming occurs. Characteristic of most mammals.

**restless leg syndrome:**  Disorder characterized by complaints of restlessness, creeping sensations, or electrical activity in the legs that prevent one from falling asleep.

**seasonal affective disorder (SAD):** A type of depression that appears to be associated with seasons of the year when the amount of daily sunlight is reduced (e.g., in North America, winter).

**sentinel hypothesis:** Theory that dreaming is a periodic arousal-security system that allows the sleeper to scan the environment for threatening signs.

**sleep apnea:** Cessation or reduction in respiration occurring during sleep. May lead to high blood pressure, enlargement of the heart, cardiac arrhythmias, and sudden death during sleep.

**sleep inertia:** A short-lived feeling of lethargy immediately following awakening from a nap.

**sleep latency:** The amount of time elapsed between trying to fall asleep and actually falling asleep.

**sleep paralysis:** Sensation of being awake but completely paralyzed. These frightening events usually occur just at sleep onset or immediately upon awakening.

**slow-drift rotation:** Technique that gradually acclimates shiftworkers to the shift rotation changes (e.g., from day to evening shift) by scheduling them to start work at a slightly later time each day until they are established in the new rotation.

**stenosis:** Narrowing of a passage. Stenosis of the upper airway can cause snoring and sleep apnea.

**subcortical brain centers:** Areas of the brain located below the cortex, such as the brain stem.

**syncope:** Brief loss of consciousness (fainting) associated with reduced cerebral blood flow.

**thermistors:** Heat sensitive electrodes, placed near the mouth and nose to monitor respiration during sleep.

**tracheostomy:** Surgical procedure that cuts a hole into the trachea (windpipe). Used to remedy life-threatening cases of obstructive sleep apnea by bypassing the site of the obstruction.

**transient insomnia:** Complaints of difficulty in sleeping that are short-lived (less than three weeks) and usually caused by a specific event.

**ultradian:** Biological rhythms that oscillate on a cycle much less than 24 hours, such as heart rate, respiration, or the REM-NREM cycle.

**uvulopalatopharyngoplasty:** Operation to widen and stiffen the throat by removing and tightening tissue in the upper airway. Used in treatment of obstructive sleep apnea and chronic snoring.

***zeitgebers:*** Time-givers. Cues from the external environment that give information about time.

# Appendix: Owl and Lark Questionnaire

If your test score indicates that you are a Moderately Morning Type, Neither Type, or a Moderately Evening Type, you are like most people. You can probably manipulate your sleep-wake cycle to adjust to weekends, shiftwork, or jet lag with only a moderate degree of difficulty. If you score as a Definitely Morning Type, you probably find that your performance peaks in the early morning and rapidly falls as evening approaches. You may have extreme difficulty in adjusting to shiftwork, jet lag, and changing schedules. If you score as an extreme owl, a Definitely Evening Type, your body is set for peak functioning at night. Adjustment to shiftwork and to jet lag is probably easier for you than for most people.

Both extremes—owls and larks—may find it difficult to stay in synch with the other members of their families, or with work schedules. Whether you are an owl or a lark, however, it is possible to reset your clock so that you fall into the normal range. Don't fight your biological clock. Either change your social and work schedules to adapt to your biological rhythm or manipulate your own biological clock by using some of the techniques discussed in this book.

Adapted from "A Self-Assessment Questionnaire to Determine Morningness-Eveningness in Human Circadian Rhythms," by J.A. Horne and O. Ostberg. *International Journal of Chronobiology*, Vol. 4, 97–110, (London: Gordon and Breach Science Publishers Ltd., 1976)

Instructions:
1. Please read each question very carefully before answering.
2. Answer ALL questions.
3. Answer questions in numerical order.
4. Each question should be answered independently of others. Do NOT go back and check your answers.
5. All questions have a selection of answers. For each question place a cross alongside ONE answer only. Some questions have a scale instead of a selection of answers. Place a cross at the appropriate point along the scale.

*Scoring*  For questions 3, 4, 5, 6, 7, 8, 9, 11, 12, 13, 14, 15, 16 and 19, the appropriate score for each response is displayed besides the answer box.

For questions 1, 2, 10 and 18, the cross made along each scale is referred to the appropriate score value range below the scale. For question 17 the most extreme cross on the right hand side is taken as the reference point and the appropriate score value range below this point is taken.

The scores are added together and the sum converted into a five-point Morningness-Eveningness scale:

|  | **Score** |
| --- | --- |
| Definitely Morning Type | 70–86 |
| Moderately Morning Type | 59–69 |
| Neither Type | 42–58 |
| Moderately Evening Type | 31–41 |
| Definitely Evening Type | 16–30 |

1. Considering only your own "feeling best" rhythm, at what time would you get up if you were entirely free to plan your day?

2. Considering only your own "feeling best" rhythm, at what time would you go to bed if you were entirely free to plan your evening?

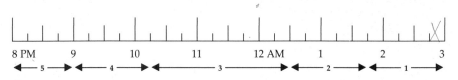

3. If there is a specific time at which you have to get up in the morning, to what extent are you dependent on being woken up by an alarm clock?

Not at all dependent.................. ☐ 4
Slightly dependent.................... ☐ 3
Fairly dependent..................... ☑ 2
Very dependent...................... ☐ 1

4. Assuming adequate environmental conditions, how easy do you find getting up in the mornings?

Not at all easy ..................... ☐ 1
Not very easy ...................... ☑ 2
Fairly easy ........................ ☐ 3
Very easy ......................... ☐ 4

5. How alert do you feel during the first half hour after having woken in the mornings?

Not at all alert ..................... ☐ 1
Slightly alert ...................... ☐ 2
Fairly alert ....................... ☑ 3
Very alert ........................ ☐ 4

6. How is your appetite during the first half-hour after having woken in the mornings?

Very poor.......................... ☑ 1
Fairly poor ........................ ☐ 2
Fairly good........................ ☐ 3
Very good......................... ☐ 4

7. During the first half-hour after having woken in the morning, how tired do you feel?

Very tired ......................... ☐ 1
Fairly tired ........................ ☐ 2
Fairly refreshed..................... ☑ 3
Very refreshed...................... ☐ 4

8. When you have no commitments the next day, at what time do you go to bed compared to your usual bedtime?

Seldom or never later . . . . . . . . . . . . . . . ☐ 4
Less than one hour later . . . . . . . . . . . . . ☐ 3
1 – 2 hours later . . . . . . . . . . . . . . . . . . . . ☑ 2 .
More than two hours later . . . . . . . . . . . ☐ 1

9. You have decided to engage in some physical exercise. A friend suggests that you do this one hour twice a week and the best time for him is between 7:00 – 8:00 AM. Bearing in mind nothing else but your own "feeling best" rhythm how do you think you would perform?

Would be in good form . . . . . . . . . . . . . . ☐ 4 *
Would be in reasonable form . . . . . . . . . . ☐ 3
Would find it difficult . . . . . . . . . . . . . . . ☐ 2
Would find it very difficult . . . . . . . . . . . ☑ 1

10. At what time in the evening do you feel tired and as a result in need of sleep?

11. You wish to be at your peak performance for a test which you know is going to be mentally exhausting and lasting for two hours. You are entirely free to plan your day and considering only your own "feeling best" rhythm which ONE of the four testing times would you choose?

8:00 – 10:00 AM . . . . . . . . . . . . . . . . . . . . ☐ 6
11:00 AM – 1:00 PM . . . . . . . . . . . . . . . . . ☐ 4
3:00 – 5:00 PM . . . . . . . . . . . . . . . . . . . . . ☐ 2
7:00 – 9:00 PM . . . . . . . . . . . . . . . . . . . . . ☑ 0

12. If you went to bed at 11:00 PM at what level of tiredness would you be?

Not at all tired . . . . . . . . . . . . . . . . . . . . . ☐ 0
A little tired . . . . . . . . . . . . . . . . . . . . . . . ☑ 2
Fairly tired . . . . . . . . . . . . . . . . . . . . . . . . ☐ 3 *
Very tired . . . . . . . . . . . . . . . . . . . . . . . . . ☐ 5

13. For some reason you have gone to bed several hours later than usual, but there is no need to get up at any particular time the next morning. Which ONE of the following events are you most likely to experience?

Will wake up at usual time and will NOT fall asleep . . . . . . . . . . . . . ☐ 4
Will wake up at usual time and will doze thereafter . . . . . . . . . . . . ☐ 3
Will wake up at usual time but will fall asleep again . . . . . . . . . . . . . ☐ 2
Will NOT wake up until later than usual . . . . . . . . . . . . . . . . . . . . ☑ 1

14. One night you have to remain awake between 4:00 – 6:00 AM in order to carry out a night watch. You have no commitments the next day. Which ONE of the following alternatives will suit you best?

Would NOT go to bed until watch was over . . . . . . . . . . . . . . . . ☑ 1
Would take a nap before and sleep after . . . . . . . . . . . . . . . ☐ 2 *
Would take a good sleep before and nap after . . . . . . . . . . . . . . . . ☐ 3
Would take ALL sleep before watch . . . . ☐ 4

15. You have to do two hours of hard physical work. You are entirely free to plan your day and considering only your own "feeling best" rhythm which ONE of the following times would you choose?

8:00 – 10:00 AM . . . . . . . . . . . . . . . . . . . . ☐ 4
11:00 AM – 1:00 PM . . . . . . . . . . . . . . . . . ☐ 3
3:00 – 5:00 PM . . . . . . . . . . . . . . . . . . . . . ☐ 2
7:00 – 9:00 PM . . . . . . . . . . . . . . . . . . . . . ☑ 1

16. You have decided to engage in hard physical exercise. A friend suggests that you do this for one hour twice a week and the best time for him is between 10:00 – 11:00 PM. Bearing in mind nothing else but your own "feeling best" rhythm how well do you think you would perform?

Would be in good form . . . . . . . . . . . . . . ☐ 1
Would be in reasonable form . . . . . . . . . ☑ 2
Would find it difficult. . . . . . . . . . . . . . . ☐ 3
Would find it very difficult . . . . . . . . . . . ☐ 4

17. Suppose that you can choose your own work hours. Assume that you worked a FIVE-hour day (including breaks) and that your job was interesting and paid by results. Which FIVE CONSECUTIVE HOURS would you select?

12  1  2  3  4  5  6  7  8  9  10 11 12  1  2  3  4  5  6  7  8  9  10 11 12
MIDNIGHT                              NOON                              MIDNIGHT
◄— 1 —►◄— 5 —► 4 ◄— 3 —►◄ 2 ►◄———— 1 ————►

18. At what time of the day do you think that you reach your "feeling best" peak?

12  1  2  3  4  5  6  7  8  9  10 11 12  1  2  3  4  5  6  7  8  9  10 11 12
MIDNIGHT                              NOON                              MIDNIGHT
◄—— 1 ——►◄ 5 ►◄ 4 ►◄——— 3 ———►◄—— 2 ——►◄ 1 ►

19. One hears about "morning" and "evening" types of people. Which ONE of these types do you consider yourself to be?

Definitely a "morning" type? . . . . . . . . . ☐ 6
Rather more a "morning" than an "evening" type . . . . . . . . . . . . . . ☐ 4
Rather more an "evening" than a "morning" type . . . . . . . . . . . . . . ☑ 2
Definitely an "evening" type . . . . . . . . . . ☐ 0

# About The Author

Richard M. Coleman is one of the nation's leading sleep experts and industrial consultants, specializing in the application of circadian principles (biological cycles) to the workplace. Formerly co-director of the Stanford University Sleep Disorders Clinic, he is currently a member of the clinical faculty at Stanford University Medical School.

Dr. Coleman divides his time between the practice of clinical psychology, in which he treats individual patients with sleep disorders, and consulting with major companies to improve round-the-clock schedules. Companies for whom he has designed schedules include Alcan, Chevron, Conoco, and Procter and Gamble. As consultant to the U.S. Olympic Committee, Dr. Coleman has studied ways of reducing jet lag and its effects on athletic performance.

As a graduate student, Dr. Coleman participated in pioneering studies conducted at the University of Chicago and at Albert Einstein College of Medicine in New York. After receiving his Ph.D. in psychology from Yeshiva University in 1979, he came to Stanford to work with Dr. William Dement. He has published numerous papers in journals such as *Science*, the *Journal of the American Medical Association*, and *Sleep*.

A frequent lecturer at scientific and professional conferences, as well as at meetings of community and industrial organizations, Dr. Coleman in 1984 led an American delegation to the Soviet Union for a Soviet-American Conference on Psychology/Sleep-Wake Disorders. His current research focuses on improving shiftwork schedules in the interest of better health and safety, both for shiftworkers and for the public whom they serve.

*Series Editor:* Miriam Miller
*Production Coordinator:* Gayle Hemenway
*Book Design:* Andrew Danish
*Graphs and Charts:* Pamela Manley

# Credits

Cover  The three-dimensional image of the eye is a hologram produced by Laser Creations of Seaside, California.

Page

4    Illustration by Pamela Manley. Adapted from M.C. Moore-Ede, F.M. Sulzman, and C.A. Fuller, *The Clocks That Time Us* (Cambridge, MA: Harvard University Press, 1982).

19   Figure 2-1. Source: Lower panels from Charles Czeisler, "Internal Organization Of Temperature, Sleep-Wake, and Neuroendocrine Rhythms Monitored in an Environment Free of Time Cues," PhD thesis, Stanford University, 1978. The alertness panel is a hypothetical adaptation of other data.

23   Illustration by Pamela Manley.

32   Photograph from Stefan Lorant, *Pittsburgh: The Story of An American City* (Lenox, MA: Authors Edition, Inc., 1964).

36   Photograph courtesy Bethlehem Steel Corp.

40   *New York Times* article copyright 1923 by the New York Times Company. Reprinted by permission.

65   Photograph courtesy Wide World Photos.

75   Figure 4-2. Source material from *War in the Falklands*, by *The Sunday Times* of London Insight Team (New York: Harper & Row, 1982).

94   Article reprinted courtesy of Reuters.

100  Photographs courtesy Ted Spagna. Copyright Ted Spagna. From the catalog of the exhibit entitled "Dreamstage: An Experimental Portrait of the Sleeping Brain."

105  Figure 5-1. Source: William C. Dement, *Some Must Watch While Some Must Sleep* (Stanford, CA: The Portable Stanford, Stanford Alumni Association, 1972).

108  Figure 5-2. Source: Anthony Kales, ed., *Sleep: Physiology and Pathology, A Symposium* (Philadelphia: J.B. Lippincott Company, 1969).

116  From James Thurber, *The Seal in the Bedroom* (New York: Harper & Row, 1932). Copyright 1932, 1960, James Thurber.

135  Illustration by Jim M'Guinness.

160  Illustration by Thomas Nast, courtesy the New York Public Library, Astor, Lenox and Tilden Foundation.

186  Photograph by Chuck Painter, Stanford News and Publications.

# Index

Body temperature, 16–17, 19, 20, 24, 26, 58–59
Brain lesions, 5–6
Brain stem, 117, 118, 119
Brain-wave measurements, 20, 89–90, 101–107, 155
Brandeis, Louis D., 37–38
Breathing, *see* Respiration abnormalities
Bronx studies, 6–11
Burlington Northern train collisions, 1–2, 21, 170

Caffeine, 81, 139, 150, 167–168
Canary Islands crash, 72–73
Cancer therapy, 27–28
Cardiovascular diseases, *see* Heart disease
Carnegie, Andrew, 35, 39
Carnegie Steel Company, 35
Carskadon, Mary, 132, 155
Cataplexy, 155–156, 157, 158
Cats, 89–90, 106
Chernobyl disaster, 31
Children: pre-adolescent, alertness of, 165; Stage 3–4 sleep of, 103
Chronic insomnia, 137, 138–141
Chronobiology, 12
Chronograms, 26
Chronotherapy, 141–143, 144
Circadian rhythms, 12, 16–17, 21–28; air travel affecting, 64, 73, 74, 76, 80, 84; during daylight saving time, 61; in infants, 5–6; naps relating to, 100; work schedules affecting, 46, 47, 48, 51–52, 59
Circannual cycles, 12, 28–29
Computer: metaphor for dreaming, 120
Concorde, 79
Cortisol, 19, 27
Crick, Francis, 120
Cushing's disease, 27
Cytarabine, 27
Czeisler, Charles, 6, 42

Dark-light cycles, 25–26, 60, 148
Day, Kerry, 56–58
Daylight saving time, 60–61
Death: from sleep apnea, 161; timing of, 28–29. *See also* Mortality rates
de Candolle, 5
de Mairan, Jean Jacques d'Ortous, 3, 12, 13
Dement, William, 24, 92, 104, 154, 155
Dementia, 162
Depression: mental, 25, 136, 138, 151, 168
de Saint-Denys, Marquis d'Hervey, 121–122
Diagnosis: insomnia, 136–141, 144, 150–51; sleep apnea, 161
Diary, sleep, 149, 150
Dickens, Charles, 158

Diet, 81–82, 147, 150
Digestive disorders, 59, 73
Diphenhydramine, 169
Diurnal mode, 31, 32, 170
Dreaming, 90, 104, 106–107, 111–129
Driving, 20–21, 167
Drugs: circadian rhythm and, 27–28; insomnia and, 136, 139, 150, 151; and jet lag, 76, 80, 81, 83–84; night terrors suppressed by, 125; sleeping aid, 168–169 (*see also* Sleeping pills)
Duhamel, Henri-Louis, 4
Dulles, John Foster, 66, 80

Economo, Von, 89
Edison, Thomas, 33, 97–98
EEG (electroencephalograph), 89, 102, 103, 155
Elderly people: insomnia of, 134; sleep apnea of, 134, 161–162; Stage 3–4 sleep of, 103
Electrosleep, 143–144
Eli Lilly Company, 50
Emotions: insomnia and, 134, 136, 138–139; narcoleptics and, 157; REM sleep helping with, 118. *See also* Anxiety; Depression, mental
Encephalitis, 89, 94
England: insomnia studies in, 133
Environment: sleep, 145–149
Erections: nocturnal, 125–126
Eskimos, 25
Europe, 50, 169. *See also individual countries.*
Exercise, 145–146, 150
Eye movements: monitoring of, 20–21, 101–107
Eyes: sleep deprivation effects on, 92

Falklands War, 75–76
Fatigue: air travel, 63–64, 65, 66, 71–72, 74, 76, 170–171; from sleep deprivation, 92
Federal Aviation Administration (FAA), 72, 74
Federal Republic of Germany, 60
Finland, 133
Firestone Tire & Rubber Company, 44
Fish, 107
Flurazepam, 168
Ford, Henry, 38
Franklin, Ben, 149
Free-running, 4, 9, 12, 21–22, 60–61, 69
Freud, Sigmund, 113–114, 115–116, 117, 119
Fruit flies (*Drosophila*), 6

Gardner, Randy, 24
Gastrointestinal disorders, 59, 73
Gatty, Harold, 64–66

German air crew scheduling, 74
German studies, 6, 20, 60, 74
Gigantocellular tegmental field (GTF), 119
Great Salt Lake Minerals and Chemicals
    Corporation (GSL), 42–44, 45–49, 54, 171
Greeks, 3, 113, 162
Growth hormone, 19, 26
Guilleminault, Christian, 127

Habits, 139–140
Halberg, Franz, 60
Halcion, 81
Hallucinations, 87, 91, 92, 118, 129
Hanford schedule, 45
Harding, W.G., 39
Health: insomnia and, 134–136, 139; and
    shiftwork, 58–61, 73; sleep apnea and,
    161; sleeping pills and, 144; and sleep
    length, 97–98, 137
Heart disease: circannual rhythms and, 29;
    insomnia and, 134; REM sleep and, 126–
    127; and shiftwork schedules, 60; and
    sleep length, 97
Heliotropes, 3–5
Herophilus of Alexandria, 3
Hilprecht, Herman, 121
Hippocrates, 3
Homestead strike, 35
Hoover, Herbert, 39
Hormones, 16, 17, 18, 19, 25, 26, 27, 28
Houston, Texas, 134
Hunter, J., 16
Hypersomnia, 89, 153–154
Hypnotics, see Sleeping pills
Hypotoxins, 88
Hypoxia, 161

Impotence, 125–126
Infants: biological clock of, 5; REM sleep of,
    107, 118; Infradian cycles, 12
Insomnia, 26, 41, 89, 94–95, 97, 131–151; in
    Arctic Circle, 25; among flight crews, 71;
    shiftworker, 59, 133; social aspects of, 143;
    on Sunday nights, 22–23
Interchurch World Movement, 37
Internal desynchronization, 24
*Interpretation of Dreams*, 113–114
Italy, 89, 100–101, 133–134

Jackson, Hughling, 118
Jacobson, Edmund, 106
Japan, 170
Jet lag, 26; air travel, 63–85; blue collar, 46,
    61
Johnson, Laverne, 134–135
Johnson, Lyndon, 80
Jones, William R., 35

Jung, C.G., 114–115, 116

Keogh, Preston, 10–11
Kinsey report, 126
Kleitman, Nathaniel, 16, 17, 104

Laboratory of Human Chronophysiology, 6–
    11
Ladd, G.T., 106
Larks and Owls, 15, 59
Leg movements: periodic, 136, 139, 151
Life cycle: chronobiology of, 28–29
Life expectancy, see Longevity
Life span: of air crews, 73–74; and shiftwork,
    59–60
Light bulbs: invention of, 33, 98
Light-dark cycles, 25–26, 60, 148
Locus cerules (LC), 119
Longevity: insomnia and, 134; and
    shiftwork, 60; sleep length and, 97, 134
L-tryptophan, 147, 150
Lucid dreams, 121
Luther, Martin, 113

Mammals, 107–8, 118. See also Animals
Melatonin secretion, 25
Memorization, 17
Memory, 95
Methylxanthine, 81
Mexico, 99, 100–101
Mice, 60
Microsleeps, 91
Miles, Laughton, 17
Milk, 147
Monday morning blues, 23
Montefiore Hospital, 6
Mortality rates: of shiftworkers, 59–60; with
    sleeping pills, 162; and sleep length, 97
Motivation, 92–93, 98, 166, 167
Multiple Sleep Latency Test (MSLT), 155,
    157, 164–168 passim
Muscle activity monitoring, 101–102, 104,
    105, 106, 107

Napping, 99–100, 150
Narcolepsy, 41, 119, 125, 155–158, 165, 167
NASA, 74
Nasal snoring, 163
National Center for Health Services
    Research, 97
National Institutes of Health, 144
National Transportation Safety Board, 2
Naval Health Research Center, San Diego,
    134–35
Nervousness: of air crews, 73
New York City studies, 6–11
Nielsen, Scott N., 49